The Biomass Assessment Handbook

The increasing importance of biomass as a renewable energy source has led to an acute need for reliable and detailed information on its assessment, consumption and supply. Responding to this need, and overcoming the lack of standardized measurement and accounting procedures, this best-selling handbook provides the reader with the skills to understand the biomass resource base and the tools to assess the resource, and explores the pros and cons of exploitation.

This new edition has been fully updated and revised with new chapters on sustainability methodologies. Topics covered include assessment methods for woody and herbaceous biomass, biomass supply and consumption, land use changes, remote sensing techniques, food security, sustainability and certification as well as vital policy issues. The book includes international case studies on techniques from measuring tree volume to transporting biomass, which help to illustrate step-by-step methods. Technical appendices offer a glossary of terms, energy units and other valuable resource data.

Frank Rosillo-Calle is an Honorary Senior Research Fellow in Biomass Energy at the Centre for Environmental Policy, Imperial College London, UK, with over 30 years' experience in biomass energy.

Peter de Groot has worked in the field of renewable energy over 30 years and has been involved in many research projects including training courses for rural entrepreneurs.

Sarah L. Hemstock is Project Team Leader for the University of the South Pacific-European Union Global Climate Change Alliance Project, based in Fiji.

Jeremy Woods is a Lecturer in Bioenergy at Imperial College London's Centre for Environmental Policy.

The Biomass Assessment Handbook

Energy for a sustainable environment

Second edition

**Edited by Frank Rosillo-Calle,
Peter de Groot, Sarah L. Hemstock
and Jeremy Woods**

Routledge
Taylor & Francis Group

LONDON AND NEW YORK

earthscan
from Routledge

First published 2006 by Earthscan
This edition published 2015
by Routledge
2 Park Square, Milton Park, Abingdon, Oxon OX14 4RN

and by Routledge
711 Third Avenue, New York, NY 10017

Routledge is an imprint of the Taylor & Francis Group, an informa business

British Library Cataloguing-in-Publication Data
A catalogue record for this book is available from the British Library

Library of Congress Cataloging in Publication Data
Library of Congress Cataloging-in-Publication Data
The biomass assessment handbook / edited by Frank Rosillo-Calle,
Peter de Groot, Sarah L. Hemstock and Jeremy Woods. — Second
edition.
pages cm
1. Biomass—Measurement. I. Rosillo Call?, Francisco, 1945-
TP339.B548 2015
333.95'39—dc23
2014047246

ISBN: 978-1-138-01964-5 (hbk)
ISBN: 978-1-138-01965-2 (pbk)
ISBN: 978-1-315-72327-3 (ebk)

Typeset in Bembo
by Swales & Willis Ltd, Exeter, Devon, UK

MIX
Paper from
responsible sources
FSC   FSC° C013056
www.fsc.org

Printed and bound in Great Britain by
TJ International Ltd, Padstow, Cornwall

Contents

Figures, tables and boxes

Figures

Tables

Boxes

The main related abbreviations

Bl	billion litres
boe	barrels of oil equivalent (42 US gal; 1 US gal = 4.55 litres)
CAI	current annual increment
CHP	combined heat and power
CRI	crop residue index
DBH	diameter at breast height
dLUC	direct land use change
FAO	Food and Agriculture Organization
FAOSTAT	FAO Statistical Division
GHG	greenhouse gas
GHV	gross heating value
GIS	geographic information system
GPS	global position system
HEI	high energy input
HHV	higher heating value
HP	horse power
IEA	International Energy Agency
iLUC	indirect land use change
IPCC	Intergovernmental Panel on Climate Change
LEI	low energy input
LHV	lower heating value
LPG	liquid petroleum gas
MAI	mean annual increment
mc	moisture content
MSW	municipal solid waste
Mtoe	million tonnes of oil equivalent
NDVI	normalized difference vegetation index
NEP	net ecosystem production
NHV	net heating value
NPP	net primary production (tonne per ha/year)
odt	oven dry tonne
odw	oven dry weight

PAR	photosynthetically active radiation
PET	potential evapo-transportation
RE	renewable energy
SPOT	Satellites pour l'Observation de la Terre (French Earth-observing satellites)
SRC	short-rotation coppice
toe	tonne of oil equivalent (= 42 GJ)
UV	ultraviolet

Contributors

Vineet V. Chandra is a PhD student in the area of bioenergy and climate change. His research interests are in energy efficiency, bioenergy, solar energy and climate change mitigation. He is also an Assistant Lecturer at the University of the South Pacific (USP) teaching mechanical and manufacturing engineering. After graduating with a Bachelor of Engineering Technology from USP, he joined HVAC industry and later became an energy consultant and carried out various energy auditing and efficiency projects in Fiji. During this time he also completed his Master of Science in Engineering. He currently resides in Fiji with his wife and two daughters.

Rocio Diaz-Chavez, PhD, is a Research Fellow at the Centre for Environmental Policy of Imperial College London. She has over 15 years experience in sustainability assessment and environmental management tools. She has worked on EU-funded projects related to biomass, biorefineries, climate change, and energy and sustainability assessment at a global level. She has worked on benchmarking standards related to bioenergy and contributed to the Global Bioenergy Partnership developing indicators. She obtained the Young Scientist on Environmental Management Award from SCOPE in 2010.

Peter de Groot, PhD (editor), has worked in the field of renewable energy for over 30 years, and has extensive experience in Africa. He has been involved in many research projects and training courses for rural entrepreneurs. He has worked as independent consultant for more than 20 years.

Sarah L. Hemstock, PhD (editor), is an author and adviser to the Alofa Tuvalu Small Is Beautiful project – recognized by UNESCO as one of its Decade of Achievement Projects. Currently she is Project Team Leader for the USP EU Global Climate Change Alliance Project, Visiting Fellow at Nottingham Trent University and Government of Tuvalu Honorary Ambassador – Officer for Environmental Science. Her work relates to practical resource management and sustainable development issues relevant to some of the world's most disadvantaged communities and vulnerable ecosystems. Her research examines community-initiated sustainable development and aid effectiveness.

Nicole Kalas is a Doctoral Researcher at Imperial College London with experience in land use modelling, life-cycle assessment (LCA) and carbon foot-printing, techno-economic and sustainability assessments of bioenergy supply chains and the bioeconomy. Her PhD research is focused on system dynamic modelling of potential pathways for the sustainable intensification of agriculture and the integrated production of food, feed, fibres, energy and feedstocks for the bioeconomy. Nicole has worked on international climate change mitigation projects as a researcher in academia and the private sector in various countries.

Rubens A.C. Lamparelli, PhD, has a first degree in Agriculture Engineering from the University of Campinas (UNICAMP), Brazil, with an MSc in Remote Sensing (National Institute for Space Research, Brazil) and a PhD in Geo-processing at the University of São Paulo University (USP). He has experience in crop monitoring and yield estimates mainly in sugar cane, soybean and coffee. He is currently Senior Researcher at the Interdisciplinary Center for Energy Planning (NIPE) at UNICAMP and supervises post-graduate students at the School of Agricultural Engineering.

Teuleala Manuella-Morris is from Nanumea, the most northerly island of Tuvalu. As head of the Department for Community Affairs for 19 years, she has vast experience on sustainable development issues relevant to women and isolated island communities. She also headed the National Agricultural Station in Vaitupu. Recently she developed the first National Social Policy, improving the well-being of vulnerable and disadvantaged Tuvaluans and initiating the payment of a state old-age pension. Currently she is In-Country Coordinator for the USP EU GCCA project, and pursuing her PhD at the University of the South Pacific.

Frank Rosillo-Calle, PhD (editor), is Honorary Senior Research Fellow in Biomass Energy at the Centre for Energy Policy, Imperial College London. He has more than 30 years' research experience in biomass for energy field and has published extensively. His main interests include biomass resource assessment, biomass for energy (bioenergy and biofuels: production, conversion and use), agriculture and food security, and social and environmental implications. He has extensive experience in international research and has taught biomass for energy at various universities.

Vikram Seebaluck, PhD, is Senior Lecturer and Ex-Head of the Department of Chemical and Environmental Engineering at the University of Mauritius. He has been a Co-Leader for an EC-funded thematic research network on sugar cane in southern Africa. He co-authored a book entitled *Bioenergy for Sustainable Development and International Competitiveness* (2012). He is a contributor to the SCOPE Bioenergy & Sustainability project and is the Regional African Focal Coordinator with the UN University Institute for Integrated Management of Material Fluxes and Resources (UNU-FLORES) which focuses on the sustainable management of environmental resources

Ranjila Singh is a part-time publications consultant for the Pacific Centre for Environment and Sustainable Development and part-time teaching assistant at the Faculty of Science Technology and Environment, University of the South Pacific. She has completed a Master of Science degree and post-graduate diploma in climate change and a Bachelor's degree in Environmental Science. Her research interests are in the area of climate change and energy.

Alexandre B. Strapasson, PhD, is a Research Associate at Imperial College London, and a Visiting Lecturer in Biofuels at IFP School in Paris. He has worked on bioenergy and complex systems in projects with the private sector and governments. Alexandre also worked for several years at the Brazilian Government as Director of the Department of Sugarcane and Agro-Energy at the Ministry of Agriculture, Livestock and Food Supply (MAPA), and as UNDP Consultant for energy and climate change affairs at the Ministry of the Environment (MMA). He is an Agricultural Engineering graduate, and has undertaken post-graduate studies at the University of São Paulo, IFP School, OECC Japan, and Imperial College London, with a focus on energy and environmental sciences.

Jansle Vieira Rocha, PhD, is an agricultural engineer with an MSc in agricultural engineering at University of Campinas (UNICAMP), Brazil, and a PhD in Applied Remote Sensing at Cranfield University, UK. He is currently Associate Professor at UNICAMP, where he leads the geo-processing/remote sensing research group (GEO) and the Geo-processing Lab of the School of Agricultural Engineering. He has worked for the European Commission Joint Research Center (JRC) on crop monitoring for food security, and for CONAB, the food supply agency of the Brazilian Ministry of Agriculture, leading the Geosafras Project (crop monitoring with remote sensing).

Lei Wang, PhD, is a Marie Curie Research Fellow (EU FP7) at Imperial College London. Her current research focuses on assessing environmental perspectives of first- and second-generation biofuels, particularly using attributional and consequential Life-Cycle Assessment approaches. She did her PhD research in bioenergy at Imperial College London, funded by the Porter Institute, on a project evaluating technology and economic feasibility, and environmental performance for bioethanol production from waste papers. She also holds an MSc in Chemical Engineering at Imperial College, and a BEng in Chemical Engineering from Tsinghua University, China.

Jeremy Woods, PhD (editor), is a Lecturer in Bioenergy at Imperial College London working on the interplay between development, land use and the sustainable use of natural resources. He is a member of the Bioeconomy Platform of Climate-KIC which is dedicated to the development of advanced bio-renewables, and the Royal Society's Leverhulme Africa Awards Assessment Panel. His research links environmental impacts with techno-economic and sustainability assessment frameworks and is applied to policy making and industry standards. He is a member of the Scientific Committee on Problems of the Environment (SCOPE).

Foreword

Energy production in the 21st century will undoubtedly be dominated by renewable energy.

Coal was the main source of energy in the 19th century gradually replaced by oil and gas in the 20th century but they are all fossil fuels, which are proving to be unsustainable for a variety reasons: exhaustion of reserves, security of supply and severe pollution problems at the local, regional and global level.

Worldwide renewables in 2013 represented 22% of global power production and 144 countries adopted renewable energy targets and policies to reach such targets. The most commonly used renewable sources (hydro, wind and photovoltaics) are all solar energies and intermittent: they require storage systems when the sun is not shining, the wind is not blowing and rivers are not flowing due to the lack of water in dry seasons.

In contrast, among the renewable energy sources biomass has the unique characteristic of storing solar energy and is more evenly distributed around the globe than fossil fuels, particularly in the developing countries.

There are problems with the use of biomass, such as the very low efficiency of cooking with burning wood on a three-stone fire. Despite that, over 2.6 billion people in developing countries will continue to rely on biomass for cooking and heating in 2030, as pointed out in recent IEA studies. The task therefore is to modernize the use of biomass and this is occurring through improved cooking stoves, advanced conversion technologies for electricity generation and biofuels.

The main objective of this handbook is to provide guidance on the methods related to biomass-based fuels as well as their energy potential, which includes a discussion of methodology assessment to record supply and consumption of biomass energy, sustainability, land use changes and remote sensing.

The handbook does not shy from a discussion of the fuel-versus-food problem which in my view has been overblown in recent years. It is not the expansion of biomass use that is the main cause of indirect land use change (iLUC). Agricultural land in use today for food production is approximately 1,500 million ha per year and is expanding at a rate of about 6 million ha per year. It is against these numbers that iLUC from modern uses of biomass has to be evaluated.

This handbook, in its second edition, will clearly prove to be invaluable to anyone working in this field.

Professor José Goldemberg
University of São Paulo

Editors' note and acknowledgements

Editors' note

The second edition of this *Biomass Assessment Handbook* has built on from the previous edition, but with much greater personal contribution from the editors and all other authors. The editors recognize that many other people have contributed to the handbook in different ways, directly or indirectly. This is particularly the case for Chapters 2 and 3, and to a lesser extent Chapters 4 and 5 where the editors have made major contributions. We would like to acknowledge all of them but they are too many to name, to whom we offer our sincere apologies. The second edition incorporates a significant new contribution (Chapters 6, 7 and 8 and various appendices and case studies) based on the personal experiences of the authors.

Acknowledgements

The editors would like to thank those institutions and individuals that made possible the first edition of this handbook. They helped to plant the seeds of the *Biomass Assessment Handbook*. The same applies to this second edition, which has benefited considerably from the previous one. We extend our gratitude to all the people who have contributed directly or indirectly to this book.

Introduction

Considerable changes have taken place since the first edition of this handbook was published in 2007. We are now witnessing fundamental changes in the whole energy sector. An example is the emergence of shale gas, which in the USA has transformed the energy industry, with major worldwide implications. Another significant change has been the rapid increase in the use of renewable energy (RE) which in 2013 represented 22% of global power production, with a total capacity of 1,560 GW, though unequally distributed (e.g. in the European Union the share of RE was 72%). At the end of 2013, 144 countries had RE targets and some kind of energy-support policies (REN21, 2014).

As for biomass for energy, four main trends can clearly be identified: (1) the growing concern with sustainable production and use of biomass resources (covered in Chapter 6); (2) the concern with land use changes (direct and indirect), assessed in Chapter 7; (3) the increasing use in combined heat and power (CHP) and district heating and cooling systems, e.g. global biomass-based electricity generation in early 2014 reached 88 GW; and (4) the rapid expansion of biofuels, which saw production increased to 116.6 billion l at the end of 2013. And, in early 2014, over 60 countries supported transport biofuels through regulatory and mandatory policies (REN21, 2014).

It is important to recognize from the outset that the use of biomass for energy is strongly linked to developments of other RE and fossil fuels. In the industrial countries, fossil fuels, especially oil, have for decades dominated the energy we consume, and our entire economy. Transportation systems have become dangerously and overwhelmingly dependent on oil. With high oil prices (though currently lower), and political upheavals, particularly in the Middle East, it is more urgent than ever to diversify away from oil and start planting the seeds of a truly fundamental change in the way we produce and use energy. In this new scenario, uncertain as it may be, bioenergy is ensured an increasingly important role.

Bioenergy is no longer a 'transitional energy source' as often portrayed in the past, as demonstrated by the increasing number of countries with policies supporting RE. Further, bioenergy can no longer be regarded as just 'the poor person's fuel' as it is increasingly recognized as an energy source that can provide the modern consumer with convenient, reliable and affordable services. The focus, therefore, should now be on the development and production of bioenergy for modern applications: a far cry from burning wood on a three-stone fire.

However, in reality for most people in the poorer developing world 'life is a struggle for biomass'. For the three-quarters of the world's population who live in developing countries biomass energy is their number one source of primary energy. Some countries, for example Burundi, Nepal, Rwanda, Sudan and Tanzania, obtain over 80% of their primary energy from biomass. Biomass is not only used for cooking in households and many institutions and service industries, but also for agricultural processing and in the manufacture of bricks, tiles, cement, fertilizers, etc. There is often a substantial use of biomass for these non-cooking purposes, especially in and around towns and cities. Unquestionably, for a large proportion of the world population biomass will continue to be the prime source of energy for the foreseeable future.

As a study by the IEA (2002) so rightly puts it, 'Over 2.6 billion people in developing countries will continue to rely on biomass for cooking and heating in 2030. . . an increase of more than 240 million. [In 2030] biomass use will still represent over half of residential energy consumption'.

Traditional and modern uses of biomass energy

Traditional biomass is readily obtained from wood, twigs, straw, dung, agricultural residues, etc. Biomass is burned directly for either heat or to generate electricity, or it can be fermented to alcohol fuels, anaerobically digested to biogas or gasified to produce high-energy gas.

Along with agricultural expansion to meet the food needs of increasing populations, the overuse of traditional biomass energy resources has led to the increasing scarcity of hand-gathered fuelwood, and to problems of deforestation and desertification. Even so, there is an enormous untapped biomass potential, particularly in the improved utilization of existing forest and other land resources, higher plant productivity and underutilized agroforestry residues and waste.

Energy efficiency

Biomass in its 'raw' form is often burned very inefficiently, so that much of the energy is wasted. For example, where fuelwood is used for cooking in rural areas using traditional methods, the per-capita use of energy is several times greater that when gaseous or liquid fuels are used. This results in a huge waste of energy resources.

Energy planners should consider the application of advanced technologies to convert raw biomass into modern, convenient energy carriers such as electricity, liquid or gaseous fuels, or processed solid fuels, in order to increase the energy that can be extracted from biomass.

Multiple uses

In addition to food and energy, biomass is a primary source of many essential daily materials. The multiple uses for biomass have historically been presented as the six 'Fs': food, feed, fuel, feedstock, fibre and fertilizer. Biomass products are frequently a source of a seventh 'F': finance. Such multiple uses are increasing the demand for biomass as demand for natural products continues to grow.

Because of the almost universal, multi-purpose dependence on biomass, it is important to understand interrelations between these many uses, and to determine the possibilities for more efficient production and wider uses in future. The success of any new form of biomass energy will depend upon the use of reasonably advanced conversion technology. Above all, it is fundamentally important to move away from food-based crops to avoid competition with food production (see Case study 9.4), particularly when it comes to biofuels. Thus the future of biomass for energy will also largely be determined by how second- and third-generation biofuels develop.

Sustainable production and use of biomass for energy

As indicated above, sustainability has become a central tenet that conditions the use of biomass for energy. And most policies, particularly in the European Union, strongly link support for biomass with conditions about its long-term sustainability, with major implications. For this reason a new chapter (Chapter 6) has been added to this handbook which covers in detail the methods for dealing with this new reality.

The sustainable production and conversion of plants and plant residues into fuels offers a significant opportunity to alleviate the pressure on forests and woodlands for use as fuel. Along with agricultural clearances, these pressures have been the major threats to forest and tree resources, wetlands, watersheds and upland ecosystems.

Biomass for energy also has an important role to play in the mitigation of climate change. Using modern energy-conversion technologies it is possible to displace fossil fuels with an equivalent biofuel. When biomass is grown sustainably for energy, with the amount grown equal to that burned for a given period, there is be no net build-up of CO_2, because the amount of CO_2 released in combustion is compensated for by that absorbed by the growing energy crop.

With ever-pressing environmental concerns over de-vegetation and deforestation, desertification and the role of CO_2 in climate change, it is crucial that we move to the sustainable and efficient use of biomass energy as both a traditional fuel and in modern, greenhouse-gas-neutral commercial applications.

Land use change management

Land use (direct and indirect) has also emerged as a key conditioning factor for bioenergy, which is still hotly debated, and for this reason the editors felt the need to assess its implications in much greater detail (see Chapter 7).

The optimum sustainable production and use of biomass is, ultimately, really a problem of land management. New and traditional agricultural agroforestry and intercropping systems maximize energy and food production. At the same time these systems diversify land use by producing ancillary benefits such as fodder, fertilizers, construction materials and medicines, while at the same time increasing environmental protection by, for example, maintaining soil fertility and structure under long-term cultivation.

Large-scale energy plantations, for example, require also that a country or region has a policy on how to use its land. Such a policy would go far beyond energy to include sections on food production and prices, land reform, food exports and imports, tourism and the environment. Lack of such integrated policies often gives rise to conflicts between use of land for fuel, food and other uses, which can result in the indiscriminate clearing of forests and savannah for agricultural expansion.

Why this handbook?

The fundamental objective of this handbook is to provide guidance on methods for assessing biomass-based fuels. However, given the intertwined nature of biomass, it is important to make the general reader aware of its energy potential and wider implications. Since the publication of the first edition, a lot of water has gone under the bridge. The second edition of the handbook has tried to incorporate all the major changes affecting the deployment of biomass-based fuels, and its pros and cons. Many methods for assessing biomass have changed little and have consequently been updated only when required.

Despite of the overriding importance of biomass energy, its role continues to be largely unrecognized. Many developing countries are currently experiencing acute shortages of biomass energy. Programmes to tackle this breakdown in the biomass system will require detailed information on the consumption and supply of biomass. Governments and other agencies need detailed knowledge of energy demand along with information on the annual yield and growing stock of biomass resources in order to plan for the future. Clearly some standardized measurements are required of the supply and demand for the various forms of biomass energy, similar to those available for fossil fuels.

There remains a general lack of information on the requirements and use of biomass energy. There are no standard measuring and accounting procedures for biomass, so it is often impossible to make comparisons between sets of existing data which, generally, is often inaccurate. This handbook is intended to provide a practical, common methodology for measuring and recording the consumption and supply of biomass energy. With the aim of appealing to the wider audience inclusion of too many technical details has been avoided.

This handbook is the result of many years of the editors' and individual contributors' personal experience. We have gathered material from our own fieldwork, teaching courses and other existing sources. We wanted to demonstrate the importance of biomass and show how to assess such important resources step by step. This handbook places particular emphasis on traditional bioenergy applications, but modern uses are also considered. This is because the traditional applications represent the most serious difficulties when it comes to measurement. It is important to note that measuring techniques, as with everything else, are not static but constantly changing.

Biomass resources are potentially the world largest and most sustainable source of energy, which in theory at least could contribute over 800 EJ without affecting the world's food supply. By comparison, current energy use is just over 450 EJ. It is not surprising that biomass features strongly in virtually all the major global energy supply scenarios. This handbook is intended to help those people interested in understanding the biomass resource base and the techniques to measure it; this is described in the various chapters as follows.

Chapter 1 presents a short overview of the biomass resource potential, current and future uses (various scenarios); technology options, traditional versus modern applications and difficulties with data quality. The aim is to familiarize the non-specialist reader with bioenergy, the importance of biomass as an energy source and its likely energy contribution to the world energy matrix in the future.

Chapter 2 introduces some of the main problems in measuring the use and supply of biomass, outlines the system for classifying biomass used in this handbook and sketches general methods for assessing biomass. It also looks, albeit briefly, at biomass flow charts, units for measuring biomass (e.g. stocking, moisture content and heating values), weight versus volume, calculating energy values, the implications for setting up a biomass energy plant and, finally, considers possible future trends. In short, this chapter summarizes all the major issues presented in this handbook.

Chapter 3 brings together various types of biomass (woody and herbaceous), planning issues, (land constraints, land use, tenure rights), climatic issues, etc. It looks in some detail at the most important methods for accurately measuring the supply of woody biomass for energy, and in particular techniques for forest mensuration, determining the weight and volume of trees, measuring the growing stock and yield of trees, measuring the height and bark and measuring the energy available from dedicated energy plantations, agro-industrial plantations and processed woody biomass (woody residues, charcoal).

Chapter 4 is concerned primarily with non-woody biomass, secondary fuels and tertiary fuels. New crops being considered as possible dedicated energy crops, e.g. *Miscanthus*, reed canary grass and switchgrass, may be included in the analysis. It also examines densified biomass (wood pellets, wood chips, torrefied wood), which is increasingly being traded internationally; looks in some detail at secondary fuels (biodiesel, biogas, ethanol, methanol and hydrogen) and briefly reviews tertiary fuels (municipal solid waste), as their development can have major impacts on biomass resources. Finally the chapter assesses animal traction, which still plays a major role in many countries around the world and also impacts on biomass resources.

Chapter 5 deals in some detail with various methods for obtaining reliable data on biomass energy consumption. It is structured to have particular relevance for the fieldworker looking at the feasibility of smaller-scale bioenergy projects. The emphasis is on community-level consumption in rural areas of developing countries, with respect to the amount and type of biomass resource consumed and that available for project activities. Chapter sections examine suitable assessment methods, appropriate analysis and the assessment of availability of appropriate resources for satisfactory formulation of a bioenergy project.

Chapter 6 assesses the sustainability issues arising from biomass-based fuels. It reviews selected tools and methodologies that can be used to assess and identify whether the different practices for biomass production and use for energy can be deemed as sustainable. Practical tools such as Environmental Impact Assessment (EIA), Social Impact Assessment (SIA) and Sustainability Assessment (SA) are discussed, as well as certification and monitoring costs.

Chapter 7, on land use assessment for sustainable biomass, deals with land use change (direct and indirect) and the various models used for assessing land use. It also provides details of the Global Calculator tool.

Chapter 8, on remote sensing – unlike in the previous edition where the focus was on forestry – examines the basic principles of remote sensing, image analysis and applications in land use mapping and monitoring. This is because, and although same changes have occurred in the forestry sector, land use changes have become a key factor in recent years. The increasing use of biomass and bioenergy is related to recent dynamics in land use, with possible impacts on environment and climate. Examples and case studies are presented, mainly for land use change related to bioenergy crops, such as sugar cane in Brazil and Africa.

Chapter 9 comprises four case studies, each dealing in detail with a particular aspect of biomass for energy. They are used to illustrate step-by-step methods for calculating biomass resources and uses in modern applications, based on fieldwork experiences, or to illustrate potentially major changes or trends in a particular area. These are: (1) biotrade, which examines the development of international biotrade in bioenergy and its wider implications, (2) biogas use in small island communities, (3) how to build, step by step, a biomass flow chart for bioenergy and (4) food-versus-fuel debate.

Chapter 10 comprises seven technical appendices to assist the reader in pursuing some technical data in greater detail. These are: (1) a glossary of related terms, (2) the most commonly used biomass symbols, (3) basic definitions of energy units, (4) conversion factors of wood, fuelwood and charcoal, (5) heating values and moisture contents, (6) measuring sugar and ethanol yields and (7) carbon contents of fossil fuels and bioenergy feedstocks.

References

IEA (2002) *Energy outlook 2000–2030*. IEA, Paris.
REN21 (2014) *Global status report 2013*. REN21 Secretariat, Paris.

1 Overview of bioenergy

Frank Rosillo-Calle and Jeremy Woods

Introduction

Bioenergy is not an energy source in transition, as it is often portrayed, but a resource that is becoming increasingly important as a modern energy carrier. Many readers will consider biomass for energy as a key component of our future energy needs but not all agree as to the extent of this role. With a very large potential role of biomass for energy also come increasing concerns about potential impacts, both positive and negative, e.g. land use, food production, biodiversity, environment, climate change, sustainable development, etc. The widespread use of biomass for energy needs to be assessed within its wider societal context, examining all the pros and cons; in this researchers have a special responsibility to look for the bigger picture, and to see biomass for energy as *part of the problem and also as part of the energy solution.*

Since the first edition of this handbook the energy sector has become even more dynamic and unpredictable, so this chapter differs significantly from that of the first edition. Take, for example, the emergence of shale gas, which has transformed the energy scene in the USA, with huge implications for global energy markets. Although this handbook focuses on the methodologies for assessing biomass for energy, bioenergy needs to be contextualized within the over-arching drivers for global energy provision, and so major changes have been made to the handbook since the previous edition.

Hence this chapter provides an overview of bioenergy, the world's largest source of renewable energy (RE), not a set of statistics. It examines the role of biomass energy scenarios and the potential of biomass (examining traditional versus modern applications and linkages between the two), it details the difficulties in compiling information and classifying biomass energy, it looks at the barriers to the use of biomass energy and finally it examines the possible future role of bioenergy.

Historical role of bioenergy

Throughout human history biomass in all its forms has been the most important source of our basic needs, often summarized as the six 'Fs': food, feed, fuel, feedstock, fibre and fertilizer. Biomass products are also frequently a source of a seventh 'F': finance. Until the early 19th century biomass was the main source of energy for industrial and non-industrial countries alike, and, indeed, still continues to provide the bulk of energy for many developing countries. Increasingly, particularly in industrial countries, it is becoming a driving force in modern and industrial applications, raising resource-use efficiencies (reducing waste for example) and increasing resilience of energy provision.

Past civilizations are the best witness to the role of bioenergy. Forests have had a decisive influence on world civilizations, which flourished as long as towns and cities were provisioned by forests and food-producing areas. Wood was the foundation on which past societies were built. Without this resource, civilization failed: forests were for them what oil is for us today (Rosillo-Calle and Hall, 1992; Hall et al., 1994). For example, the Romans used enormous quantities of wood for building, heating and all sorts of industries. The Romans commissioned ships to bring wood from as far away as France, North Africa and Spain. The need for wood as the material for architecture and shipbuilding, and the fuel for metallurgy, cooking, cremation and heating, left Crete, Cyprus, Mycenaean Greece and many areas around Rome bereft of much of their forests (Perlin and Jordan, 1983). When forests were exhausted, these civilizations began to decline.

The first steps towards industrialization were also based on biomass resources. Take charcoal, for example, used in iron smelting for thousands of years. Archaeologists have suggested that charcoal-based iron making was responsible for large-scale deforestation near Lake Victoria, Central Africa, about 2,500 years ago. In modern times, Addis Ababa is a good example of dependency on woodfuel. Ethiopia did not have a modern capital until the establishment of modern eucalyptus plantations, which early in the 20th century allowed the government to remain in Addis Ababa on a continuous basis. Before a sustainable source of biomass was secured the government was forced to move from region to region as the resources became exhausted (Hall and Overend, 1987).

In fact, some historians have argued that North America and Europe would not have developed without abundant wood supplies, since the Industrial Revolution was initially only possible due to availability of biomass resources. Britain is an excellent example which, thanks largely to its forests, was able to become one of the world's most powerful countries. Initially, forests, mainly of oak, covered two-thirds of Britain. The wood and charcoal produced from these forests was the basis for the Industrial Revolution, and continued to fuel industrial development in Britain until well into the 19th century (Schubert, 1957).

Worldwide, biomass fuels are used for cooking, heating and lighting in households and many institutions and cottage industries, ranging from brick and tile making, metalworking, bakeries, food processing and weaving to restaurants, and so forth. More recently, many new plants are being set up to provide energy from biomass directly through combustion to generate electricity or heat, or as combined heat and power (CHP) facilities. Contrary to the general view, biomass utilization worldwide remains steady or is growing for three broad reasons:

- population growth,
- increasing demand for energy,
- increasing environmental concerns.

The current role of bioenergy

Today, biomass continues to be the main source of energy in many developing nations, particularly in its traditional forms, currently providing nearly 70 EJ of primary energy globally. On average biomass provides 35% of the energy needs of three-quarters of the world's population, rising to between 60 and 80% or more in the poorest developing countries. However, modern biomass energy applications are increasing rapidly in both industrial and developing countries, so that they now account for 30–35% of total biomass energy use. For example, the USA obtains about 4% and Finland and Sweden 25% of their primary energy from biomass. Table 1.1 provides an overview of biomass for energy supply.

Biomass is a fuel that will continue to be the prime source of energy for many people for the foreseeable future. For example, an International Energy Agency (IEA) (2002) study, still valid today, concluded:

Table 1.1 Bioenergy supply, feedstocks and associated land demand estimates for 2010 (Woods et al., 2014)

	Global production (EJ)	Feedstock	Land occupied (million ha)
Global primary energy	520	Predominantly fossil	Not quantified
Total bioenergy	62	All forms, traditional and modern	≈50
Traditional bioenergy	40	Mostly from residues, wastes and harvesting parts of live trees (pollarding)	Not quantified
Modern bioenergy	21.5		≈50
Biofuels	4.2	Agricultural crops	<13
Heating (domestic and industrial)	13	2/3 residues and wastes, 1/3 energy crops (lignocellulosic)	≈30
Electricity	4.1	50% from energy crops, 50% from residues and wastes	≈10

Notes: derived from authors' own calculations based on IEA (2010, 2011a, 2011b, 2012a, 2012b) data. Biofuels (aggregate of national production data for 2010) from F.O. Lichts Interactive Data (2013; http://statistics.fo-licht.com/). Traditional bioenergy data derived from IEA (2011b) and the Intergovernmental Panel on Climate Change (IPCC's) Special Report on Renewable Energy Sources and Climate Change Mitigation (SRREN) (Source: Chum et al., 2011).

Over 2.6 billion people in developing countries will continue to rely on biomass for cooking and heating in 2030 (. . .) this is an increase of more than 240 million from current use. In 2030 biomass use will still represent over half of residential energy consumption. . . .

Because of the almost universal multi-purpose dependence on biomass, it is important to understand the interrelations between these many uses, and to determine the possibilities for more efficient production in future. The success of any new form of biomass energy will depend upon the availability of large-scale sustainable feedstocks and advanced processing and conversion technologies. Indeed, if bioenergy is to have a long-term future, it must be able to provide what people want: affordable, clean and efficient energy forms such as electricity, heat and liquid and gaseous fuels. This also entails direct competition with other energy sources but may also increase the resilience of supply and dampen price volatility in energy markets.

Biomass potential/scenarios

Biomass features strongly in virtually all the major global energy supply scenarios, as biomass resources are potentially the world's largest and most sustainable energy source. Biomass is in theory an infinitely renewable resource which in total comprises 220 billion oven dry tonnes (odt), or about 4,500 EJ, of annual primary production. While highly controversial, as a result of competing uses for biomass and long-term impacts on biological carbon stocks (Hall and Rao, 1999) estimated the gross theoretical annual bioenergy potential to be about 2,900 EJ (approx. 1,700 EJ from forests, 850 EJ from grasslands and 350 EJ from agricultural areas).

Bioenergy featured strongly in the latest assessments of the IPCC, with its Working Group III's 5th Assessment (AR5) report (IPCC, 2014) concluding:

> Bioenergy deployment offers significant potential for climate change mitigation, but also carries considerable risks (medium evidence, medium agreement). The IPCC's Special Report on Renewable Energy Sources and Climate Change Mitigation (SRREN) suggested potential bioenergy deployment levels to be between 100–300 EJ. This assessment agrees on a technical bioenergy potential of around 100 EJ (medium evidence, high agreement), and possibly 300 EJ and higher (limited evidence, low agreement). Integrated models project between 15–245 EJ/year deployments in 2050, excluding traditional bioenergy.

There are large variations in the many attempts to quantify the practical potential for bioenergy (see for example Slade et al., 2014 for a meta-analysis of the major global assessments of bioenergy potentials). This is due to the complex nature of biomass production and use, including such factors as the difficulties in estimating resource availability due to variability in climate, management and markets, long-term sustainable productivity and the economics of production and use. In addition, there is a large range of potential bioenergy feedstocks and crops, and conversion technologies, as well as ecological, social, cultural and environmental factors that need to be considered. Estimating biomass energy use is also problematic due to the range of biomass energy end uses and supply chains and the competing uses of biomass resources.

Further uncertainty exists about estimates of the potential role of dedicated energy forestry/crops and the traditional sources of biomass they could replace (as residues from agriculture, forestry and other sources including for example food wastes) and which have much lower and variable moisture and energy contents. Furthermore, the availability of energy sources, including biomass, varies greatly according to the level of socio-economic development. All these factors make it very difficult to extrapolate bioenergy potentials, in particular at a global scale, which is further complicated by lack of long-term data, as explained in Figure 1.1.

As stated, virtually all global energy scenarios include bioenergy as a major energy carrier in the future, as illustrated in Figure 1.1. These figures are relatively simple gross estimates of future global energy needs and the determination of the related primary energy mix. In order to achieve realistic scenarios for biomass energy use and its role in satisfying future energy demand and environmental constraints, it is important to reconcile 'top-down' and 'bottom-up' modelling approaches (see Figure 1.2).

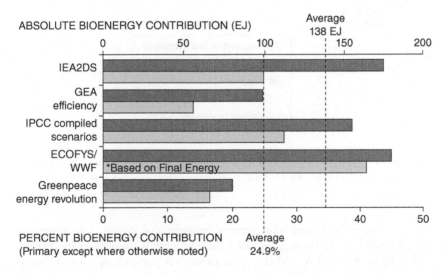

Figure 1.1 Bioenergy contribution in 2050: comparison of five low-carbon energy scenarios.

Source: Reprinted (adapted) with permission from Dale et al. (2014). Copyright 2014 © American Chemical Society.

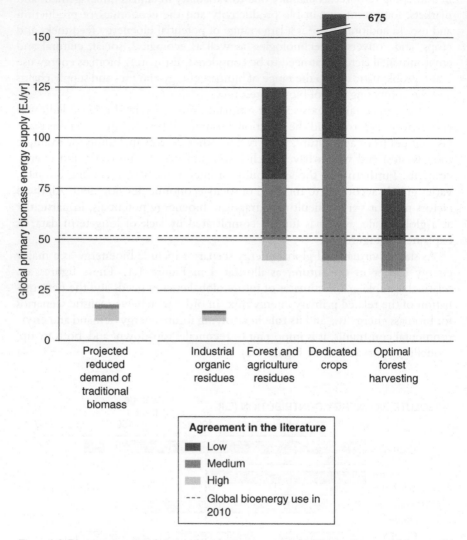

Figure 1.2 Bioenergy potentials (ranges based on expert opinion). The medium-to-high potential range is 100–300 EJ (IPCC, 2014).

Energy demand (in particular primary commercial energy consumption) increased rapidly at an average rate of 1.5% per year from 490 EJ in 2006 to over 520 EJ in 2010 (IEA, 2011b). This growth is particularly acute in some developing countries (DLGs) such as China and India, with the IEA predicting that global energy demand will increase by one-third from 2010 to 2035, with China and India accounting for 50% of the growth (IEA, 2011b). Many other

DLGs have also shown a rapid increase in demand for energy over the same period and the trend is expected to continue (IEA, 2011b). Alarmingly from a climate change perspective, nearly 50% of the additional energy needed globally between 2000 and 2010 was provided by coal (IEA, 2011b). This poses serious energy supply problems to meet such growing energy demand.

In many industrial countries significant areas of cropland are becoming available in response to pressure to reduce agricultural surpluses, particularly in the European Union and USA. For these countries there is a pressing need to find alternative economic opportunities for the land and associated rural populations; biomass energy systems could provide such opportunities, in some, although not all developing countries. There is an important strategic element in developing a biomass energy industry that needs to address the introduction of suitable crops, logistics and conversion technologies. This may involve a transition over time to more efficient crops and conversion technologies.

Residues versus energy plantations

Residues (all sources) are currently at the heart of bioenergy provision and estimates of future potentials. Residues are a large and under-exploited potential energy resource, almost always underestimated, and represent many opportunities for better utilization. There have been many attempts to calculate this potential but it remains difficult and uncertain to quantify for the reasons discussed above. In the past few years a consensus has emerged according to which, together with the second- and third-generation biofuels, residues are now considered a critical component in bioenergy's future. Large-scale energy plantations are unlikely because growing concerns with land competition, for food and biofuels (see Case study 9.4), food prices, etc.

Table 1.2 shows that the energy potentially available from crop, forestry and animal residues globally is thought to be between 55 and 111 EJ/year. However, there is considerable uncertainty and annual variability in these estimates, which remain controversial. Woods and Hall (1994) originally estimated that residue generation varied from around 3.5 to 4.2 Gt/year for agriculture and 800–900 Mt/year for forestry residues. Their estimates were based on the energy content of potentially harvestable residues based on residue production coefficients applied to Food and Agriculture Organization (FAO) Statistical Division (FAOSTAT) data on primary crop and animal production. Forestry residues are calculated from FAOSTAT 'roundwood' and 'fuelwood and charcoal' production data again using standard residue production coefficients (see Woods and Hall, 1994, for more details, and Appendix 2.1). More recently considerable efforts have been made to reduce the uncertainty in estimating the potential from residues and wastes (Table 1.2) and consensus is emerging, albeit around a very wide range (Chum et al., 2011; Creutzig et al., 2014; IPCC, 2014).

Table 1.2 Supply potentials of biomass energy resource categories by 2050

Biomass resource category		Supply potential (2050), EJ/year
1 **Forest residues**: residues from silvicultural thinning and logging; wood processing residues such as sawdust, bark and black liquor; dead wood from natural disturbances, such as storms and insect outbreaks (irregular source). Residue removal rates need to be controlled considering local ecosystem including biodiversity, climate, topography and soil factors. iLUC affects are mostly negligible but may arise if earlier uses are displaced or if soil productivity losses require compensating production. There is a near-term trade-off in that organic matter retains organic C for longer if they are left on the ground instead of being used for energy.	ASIA	3–7
	LAM	1–4
	MAF	1–3
	OECD90	7–20
	REF	2–5
	Global	17–35
2 **Unutilized forest growth**: the part of sustainable harvest levels (often set equal to net annual increment) in forests judged as being available for wood extraction, which is above the projected biomass demand for producing other forest products. Includes both biomass suitable for, e.g., pulp and paper production and biomass that is not traditionally used. The resource potential and mitigation benefit depend (besides fossil C displacement efficiency) on both environmental and socio-economic factors: the change in forest management and harvesting regimes due to bioenergy demand depends on forest ownership and the structure of the associated forest industry; and the forest productivity and C stock response to changes in forest management and harvesting depend on the character of the forest ecosystem, as shaped by historic forest management and events such as fires, storms and insect outbreaks.	ASIA	1
	LAM	22
	MAF	2
	OECD90	33
	REF	7
	Global	64–74
Forest biomass (sum of categories 1 and 2)	SRREN	0–110
3 **Agriculture residues**: manure (given separately in parentheses and not included in the agriculture residue potential); harvest residues (e.g. straw); processing residues (e.g. rice husks from rice milling). Similar environmental restrictions on harvest residue removal as for forests. iLUC effects and timing of C flows also similar, although the longer-term soil C trade-off may be less than previously believed. Residues have varying collection and processing costs (in both agriculture and forestry) depending on quality and how dispersed they are, with secondary residues often having the benefits of not being dispersed and having relatively constant quality. Densification and storage technologies would enable cost-effective collections over larger areas.	ASIA	(14) 7–30
	LAM	(8) 2–11
	MAF	(6) 3–15
	OECD90	(9) 7–13
	REF	(3) 3
	Global	(40) 28–59
	SRREN	(5–50) 15–70

4 Dedicated biomass plantations: including annual (cereals, oil and sugar crops) and perennial plants (e.g. switchgrass, *Miscanthus*) and tree plantations (both coppice and single-stem plantations, e.g. willow, poplar, eucalyptus, pine). Higher-end estimates presume favourable agriculture development concerning land use efficiency – especially for livestock production – releasing agriculture lands for bioenergy. Diets are a critical determinant, given the large land requirements to support livestock production. Large areas presently under forests are biophysically suitable for bioenergy plantations but such lands are commonly not considered available due to greenhouse gas (GHG), biodiversity and other impacts. Grasslands and marginal/degraded lands (uncertain extent and suitability) are commonly considered as available for bioenergy, but their use requires careful planning and crop selection to avoid negative impacts concerning GHG balances, water availability, biodiversity and subsistence farming and equity.

Region	Value
ASIA	6–144
LAM	7–120
MAF	4–152
OECD90	6–140
REF	3–136
Global	26–675
SRREN	*0–700*

5 Organic wastes: waste from households and restaurants, discarded wood products such as paper and demolition wood, and wastewaters suitable for anaerobic biogas production. Organic waste may be dispersed and also heterogeneous in quality but the health and environmental gains from collection and proper management through combustion or anaerobic digestion can be significant. Also must consider whether the waste had an alternative use that will need to be met from some other source.

Region	Value
ASIA	4–7
LAM	2–3
MAF	2–3
OECD90	2–3
REF	0–1
Global	10–17
SRREN	*5–>50*

Total potential: the total global and regional potentials are obtained by summing the average potentials for the different categories and then rounding to the nearest 10. This approach implies averaging over a wide range of contrasting assumptions made in the different studies concerning the development for critical parameters discussed in the text. This approach to derive global numbers is sensitive to selection of studies in the sense that including several studies using similar assumptions and models and reaching similar (high or low) potential estimates results in higher weight for these studies. However, the judgement is that the selection includes a representative mix of studies concerning models and modeller views on critical parameters.

Region	Value
ASIA	100
LAM	100
MAF	70
OECD90	140
REF	80
Global	500
SRREN	*<50–>1,000*

Source: Creutzig et al. (2014).

Notes: LAM, Latin America; MAF, Middle East and Africa; OECD90, OECD countries in 1990; REF, remaining countries; iLUC, indirect land use change. According to Creutzig et al. (2014), potentials are obtained from studies that project biomass availability for energy based on a 'food/fiber first principle' and various restrictions with reference to resource limitations and environmental concerns – but without any cost consideration – to obtain 'technical' potentials, as restricted by some sustainability considerations. The global ranges reported in SRREN (Chum et al., 2011), italic numbers in the table, have high upper limits due to a broader set of studies but have fewer restrictions and start with zero for no resource availability for energy.

In the forestry sector, and particularly commercial forestry, most residues are used to generate energy in the forest products industry, i.e. pulp and paper, particle board, construction timber, etc. But increasingly surpluses are being converted to pellets (see Appendix 4.2). This practice of using forest residues to provide power for the industry is increasing around the world. On the other hand, in countries such as China, currently undergoing a rapid fuel switch, some agricultural residues such as straw are left to rot in the fields or simply burned, while in other countries such as the UK they are utilized in modern combustion plants (see also Chapter 4). Globally, about 40% of the potentially available residues are associated with forestry and wood-processing industries; about 40% are agricultural residues, e.g. straw, sugar cane residues, rice husks and cotton residues, and about are 15–20% organic residues including animal manure.

One important reason for the development of residue-based bioenergy industries is that the feedstock costs are often low or even negative where a 'tipping fee' is levied. There are a number of important factors that need to be addressed when considering the use of residues for energy. Firstly, there may be other important alternative uses, e.g. animal feed, erosion control, use as animal bedding, as fertilizer (dung), etc. Secondly, it is not clear just what quantity of residues is available, as there is no agreed common methodology for determining what is and what is not a 'recoverable residue'. Consequently, estimates of residues often vary by a factor of five.

Energy forestry crops

The production of energy forestry crops is intensive in its land use requirement. Energy forestry crops can be produced in two main ways:

1   as dedicated plantations in land specifically devoted to this end, or
2   intercropping with non-energy forestry crops.

The future role of energy plantations is difficult to predict, since this will depend on many interrelated factors including land availability, costs and the existence of other alternatives. Estimates of land requirements have ranged from about 100 Mha to over one billion ha, with enormous variations from region to region around the world (see Woods et al., 2014 for a detailed assessment of global land availability for bioenergy). However, the potential for energy plantations could be quite limited in practice and confined to certain crops such as sugar cane. However the recent development of cellulose-based ethanol production, for example, if it becomes commercially viable, could alter considerably this panorama, although the emphasis would remain on the use of all kinds of woody residues (Janssen et al., 2013).

There remains considerable uncertainty as to the real potential of this alternative. The technical potential for energy provision from dedicated 'energy crops', including short-, medium- and long-rotation forestry, and some herbaceous crops such as sugar cane, *Miscanthus*, etc., remains unclear. Overall, it seems that the predictions of very large-scale energy plantation are unlikely to materialize, for the following reasons:

- degraded land is less attractive than good-quality land due to higher costs and lower productivity, although the importance of bringing degraded land into productive use is well recognized,
- increasing concerns with land use, food prices, population growth and political constraints,
- capital and financial constraints, particularly in developing countries,
- cultural practices and perceived moral ethical concerns with food versus fuel production,
- the need for productivity to increase far beyond what may be realistically possible, although large increases are possible,
- increasing desertification problems, and the impacts of climate change in agriculture which presently are too unpredictable,
- emergence of other energy alternatives (e.g. shale gas, wind power, solar, etc.).

Thus, the fundamental problem is not only availability of biomass resources but sustainable management and the competitive and affordable delivery of modern energy services. This implies that all aspects of both production and use of bioenergy must be modernized and, most importantly, maintained on a sustainable and long-term basis.

Biomass fuels also have an increasingly important role to play in the welfare of the global environment if used sustainably. With modern energy-conversion technologies it is possible to displace fossil fuels with an equivalent biofuel. When biomass is grown sustainably for energy there is no net build-up of CO_2, assuming that the amount grown is equal to that burned, as the CO_2 released in combustion is compensated for by that absorbed by the growing energy crop, although this notion is being constantly questioned. The sustainable production of biomass is therefore an important practical approach to environmental protection and longer-term issues such as reforestation and re-vegetation of degraded lands and in mitigating global warming.

Indeed, a combination of environmental considerations, social factors, the need to find new alternative sources of energy, political necessities and rapidly evolving technologies are opening up new opportunities for meeting the energy needs from bioenergy in an increasingly environmentally conscious world.

Continuing difficulties with data/classification of bioenergy

Information on the production and use of bioenergy is plagued with difficulties due to a lack of reliable long-term data. Even when it is available this data is often inaccurate and too site-specific. Biomass energy, particularly in its traditional forms, is difficult to quantify because there are no agreed standard units for measuring and quantifying various forms of biomass, making it difficult to compare data between sites. Furthermore, as traditional uses of biomass are an integral part of the informal economy, in most cases it never enters official statistics. Given the nature of the biomass resource, developing and maintaining a large bioenergy databank is very costly. Biomass is generally regarded as a low-status fuel, as the 'poor person's fuel'. Traditional bioenergies, e.g. fuelwood, charcoal, animal dung and crop residues, are also often associated with increasing scarcity of hand-gathered fuelwood, and have also unjustifiably been linked to deforestation and desertification.

Despite the overwhelming importance of biomass energy in many developing countries, and its increasing importance in some industrial countries, the planning, management, production, distribution and use of biomass hardly receives adequate attention among policy makers and energy planners. When relevant policy provisions are made, they are hardly ever put into practice due to a combination of factors such as budgetary constraints, lack of human resources, the low priority assigned to biomass, a lack of data, etc. Even today some countries fail to produce adequate and reliable data on bioenergy; or, worse, they fail to come up with any meaningful data at all. Considerably more long-term reliable data on all aspects of biomass production and use are still required.

However, despite this, important improvements have been made, particularly when it comes to modern applications; e.g. Austria and Scandinavian countries have good and reliable biomass energy databases. Also, in recent years, thanks to considerable efforts by some international agencies (i.e. FAO, IEA and national governments), data have improved significantly, particularly in industrial countries, although in poorer countries a lack of good biomass data often remains a serious problem. This is particularly so with economic data, which is not readily available, or is quoted in a way that makes comparisons very difficult. Far more biomass energy flow charts are needed since these can provide one of the most accurate methods of representing bioenergy data (see Case study 9.3)

Overall, the inability to fully address the indigenous biomass resource capability and its likely contribution to energy and development is still a serious constraint to the full realization of this energy potential. And despite the overriding importance of biomass energy its role is still not fully recognized. This, together with a serious lack of information, is preventing policy makers and planners from formulating satisfactory sustainable energy policies.

A further constraint is unresolved confusion with the terminology. The FAO has been trying to address this problem for years, and after many consultations important progress has been made (see Figure 1.3), going some way to solving some of these problems (see FAO/WE, 2003). Programmes to tackle this breakdown in the biomass system will require detailed information on the consumption and supply of biomass, e.g. annual yield and growing stock of biomass resources, in order to plan for the future. Clearly, some standardized measurements are required to put biomass energy on a comparable basis with fossil fuels, despite its diversity. Hopefully, this handbook will make some contributions by addressing these issues.

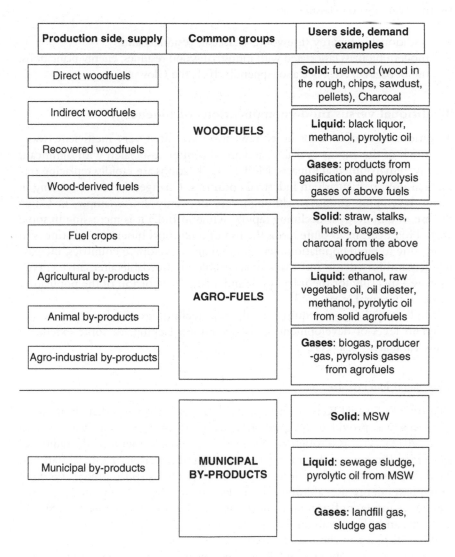

Production side, supply	Common groups	Users side, demand examples
Direct woodfuels	WOODFUELS	**Solid**: fuelwood (wood in the rough, chips, sawdust, pellets), Charcoal
Indirect woodfuels		**Liquid**: black liquor, methanol, pyrolytic oil
Recovered woodfuels		
Wood-derived fuels		**Gases**: products from gasification and pyrolysis gases of above fuels
Fuel crops	AGRO-FUELS	**Solid**: straw, stalks, husks, bagasse, charcoal from the above woodfuels
Agricultural by-products		**Liquid**: ethanol, raw vegetable oil, oil diester, methanol, pyrolytic oil from solid agrofuels
Animal by-products		
Agro-industrial by-products		**Gases**: biogas, producer -gas, pyrolysis gases from agrofuels
	MUNICIPAL BY-PRODUCTS	**Solid**: MSW
Municipal by-products		**Liquid**: sewage sludge, pyrolytic oil from MSW
		Gases: landfill gas, sludge gas

Figure 1.3 Biofuel classification scheme. MSW, municipal solid waste.

Source: Food and Agriculture Organization of the United Nations (FAO), 2004, Forestry Department, UBET: Unified Bioenergy Terminology, www.fao.org/docrep/007/j4504e/j4504e07.htm. Reproduced with permission.

The FAO classifies bioenergy into three main groups:

1   woodfuels,
2   agro–fuels and
3   urban waste–based fuels (Figure 1.3).

Biomass can also be classified as:

- traditional bioenergy (firewood, charcoal, residues) and
- modern biomass (associated with industrial wood residues, energy plantations, use of bagasse, etc.). See also Appendix 10.1, the Glossary.

Traditional versus modern applications of bioenergy

Traditional uses of biomass in its 'raw' form are often very inefficient, wasting much of the energy available, and are also often associated with significant negative environmental impacts. Modern applications are rapidly replacing traditional uses, particularly in industrial countries. Changes are also occurring in many developing countries, although very unevenly. For example, in China the use of bioenergy is declining rapidly while in India it is increasing in some areas. However, in absolute terms the use of traditional bioenergy continues to grow, due to rapid population increases in many developing countries, increasing demand for energy and a lack of accessible or affordable alternative energy sources, especially for the urban poor and a large proportion of the population living in the rural areas, etc.[1]

Modern applications require capital, skills, technology, market structure and a certain level of development, all of which are lacking, to some extent, in many rural areas of DLGs.

Traditional biomass use

The IPCC's Special Report on Renewable Energy estimates that about 40 EJ of energy was provided using traditional sources of biomass in 2010 (Chum et al., 2011). This is a rough estimate since, as already mentioned, traditional uses are at the core of the informal economy and never enter official statistics. Traditional sources of bioenergy in the poorest DLGs still represent the bulk of biomass energy use, although this is declining as many economies modernize. For example, in Burundi, Ethiopia, Mozambique, Nepal, Rwanda, Sudan, Tanzania and Uganda about 70–80% of energy originates from biomass.

The efficiency in the use of traditional bioenergy in DLGs varies considerably from between about 2 and 20%. This compares to 65–80% (or even 90%) using modern technologies, particularly in industrial countries such as Austria, Finland and Sweden (Rosillo-Calle, 2006). Hence the need to modernize all biomass energy applications must be a key requirement.

Proportionally, the general trend in modern bioenergy applications is increasing use in industrial countries and a decrease in DLGs, with important differences. In the industrial countries the increase is mostly the result of deliberate policies to support renewable energies, and a lower growth in energy demand. In the case of DLGs the increase reflects rapid energy demand at all levels triggered by economic development and low energy consumption in comparison to industrial countries, and the desire to move away from traditional bioenergy.

A study by Bhattacharya et al. (1999) shows there is an enormous potential to improve energy efficiency just through small changes. For example, the authors identified a potential saving of 328 Mt of biomass fuels in eight selected Asian countries alone. In addition, replacing all traditional stoves with improved stoves in those countries would save a further 296 Mt annually; savings in firewood in the domestic sector were estimated at about 152 Mt or 43% of the fuelwood used in households (Bhattacharya, 2004). Should this happen effectively, the IPCC estimates that about 10 EJ of modern bioenergy could become available from the more efficient use of biomass from traditional sources (IPCC, 2014). These figures give an indication of the enormous inefficiency in traditional bioenergy applications and the urgent need to introduce improved technologies.

Modern applications

The modernization of biomass embraces a large range of technologies that include combustion, gasification, pyrolysis and torrefaction, primarily for:

- household applications, e.g. small boilers, improved cooking stoves, use of biogas;
- small cottage industrial applications, e.g. brick-making, bakeries, ceramics, tobacco curing, etc.;
- large industrial applications, e.g. co-firing, CHP, electricity generation, heating, etc. (Rosillo-Calle, 2006).

Installed biomass-based electricity generation was over 76 GW worldwide in 2012 (see Table 1.3) but this is growing rapidly, with Europe being a hotspot for new electricity generation, particularly from wood pellets in coal co-generation plants in the UK, the Netherlands and France (see below).

Table 1.3 Summary of the global biomass to electricity market 2012

	Installed capacity 2012 (GW)	Growth rate 2011–2012 (%)	Operating time (h)	Estimated electricity generation in 2012 (TWh)
Solid biomass	50	3.5	3,500–7,000	175–350
Biogas	14	10	3,500–7,000	47–95
Municipal solid waste	10	5	3,500–7,000	36–72
Liquid biofuels	2	20	3,500–7,000	6–12
Total biomass	76	5	3,500–7,000	265–529

Source: Renewable Energy Focus (2013) Renewable power generation 2012, July/August edition (www.renewableenergyfocus.com/view/35689/renewable-power-generation-2012-figures/).

Until recently, transport had been one of the most rapidly increasing applications for bioenergy using, e.g., ethanol, and biodiesel, which are the best alternatives to petrol and diesel in the short to medium term, either blended or neat. Ethanol is particularly promising, with an estimated world production in 2013 of about 88 billion litres (Bl). This could easily reach 100 Bl before the end of the decade given the renewed and increasing worldwide interest, particularly in lignocellulosic bioethanol with over 30 countries in the process of implementing or planning for the use of ethanol for fuel. Biodiesel production also increased rapidly between 2000 and 2010 from just under 1 Bl to over 15 Bl in 2010 (see Woods et al., 2014). Other estimates indicate that by 2020 ethanol could provide between 3–6% of petrol, approx. 129 Bl (80 Bl gasoline equivalent) or even as much as 10%.

The development of flex-fuel (or multi-fuel) engines is part of a new fuel trend in the transport sector. This technology does not represent any revolutionary or fundamental change, but a long chain of small improvements with potentially major impacts. In a very short time this technology has been revolutionized, with improvements being added continuously. It is particularly significant that these innovations have been implemented at low cost. There is no doubt that technology will improve significantly in the near future and that current engine and fuel difficulties will be solved satisfactorily. The flexibility of flex-fuel engines (that is an engine that can run on both petrol and ethanol in variable blends) is evident, with its sophisticated system allowing the simultaneous utilization of different fuels blended in any proportion, and represents new opportunities and challenges for the industry, consumers and society as a whole, as in the case of Brazil (Rosillo-Calle and Walter, 2006).

Linkages between traditional and modern applications

It is difficult to predict just how long the shift from traditional to modern and efficient applications of bioenergy will take, or the exact technologies that will be used, given the many variable and complex factors involved, many of which are not directly related to energy. It is clear, however, that there is a long way to go and that this shift will be uneven (geographically, technically, socially, etc.) because the differences in the degree of development and the strength of concerns about environmental sustainability. This is particularly the case among developing nations, which differ so widely in their levels of socio-economic and technological development, not to mention natural resource endowment. Furthermore, the many forms of bioenergy (solid, liquid, gaseous) and the diversity of sectors involved and the many different applications will ensure that the transfer to modern bioenergy technologies will be uneven and complex.

In many parts of the world many households and cottage industries already use both biomass and fossil fuels, depending on availability and price. Reliability, social status and convenience also play an important part in the choice of energy. What is important, however, is that bioenergy can increasingly be associated with modernization and environmental sustainability.

An encouraging trend is the growth in international trade in bioenergy, which until very recently took place mostly at the local level (see Case study 9.1). Biotrade could bring many benefits, since it will increase competition and bring new opportunities to rural communities that have substantial natural resources and are close enough to good transportation networks.

Barriers

The large-scale utilization of bioenergy still faces many barriers, ranging from socio-economic, cultural, institutional to technical and trade. These barriers have been extensively investigated in the literature; e.g. Sims (2002), in his book *The Brilliance of Bioenergy*, identified a number of barriers and public concerns that need to be overcome, including:

- the possible destruction of native forests due to increasing commercial applications of bioenergy,
- the perceived problems with dioxins,
- the possible deleterious effects of large-scale dedicated energy plantations on water resources,
- possible effects on soils by continuous removal of residues for energy,
- potential monoculture problems posed by large-scale energy plantations, e.g. effects on biodiversity,
- possible effect of transporting large quantities of biomass (increased traffic),
- the perceived view of competition for land between food and fuel crops,
- to the problem of maintaining sustainable high productivity for very long periods.

These barriers should be taken into account when assessing bioenergy, and are discussed further elsewhere in this handbook.

Technology options

Many studies have demonstrated that even minor technology improvements could increase the efficiency of biomass energy production significantly, maintain high productivity of biomass plantations on a sustainable basis and mitigate environmental and health problems associated with biomass production and use. The main technology options are summarized in Table 1.4.

Following is a brief discussion of the main conversion technologies.

Combustion

Combustion technologies produce about 90% of their energy from biomass, converting biomass fuels into several forms of useful energy, e.g. hot air, hot water, steam and electricity. Commercial and industrial combustion plants can burn many types of biomass ranging from woody biomass to municipal solid waste. The simplest combustion technology is a furnace that burns the biomass

Table 1.4 Summary of the main biomass-based technological applications

Biomass technology combination	Main countries of application	Typical efficiency	Efficiency improvements	Possible policy options
Residential heating	SE, FI, DK, DE, IT, AT	20–90% thermal	Heat distribution improvement Air excess reduction (control) Market penetration	Labelling Minimum efficiency standard Investment subsidy
Large-scale power and CHP	FI, DK, SE, EE, LV, LT, AT	10–30% electrical	Large-scale heat utilization Improved heat recovery by ORC/flue gas condenser	Bonus system for efficiency improvement Minimum efficiency standard
Co-firing	DE, FI, UK, NL	35–43% electric	Heat utilization (Improve impact: increase market penetration)	No specific efficiency-related policies Efficiency already high, possibilities for stimulating co-firing in general
Waste incineration	DE, NL, DK, SE, LU	15–30% electrical	Higher steam pressures Corrosion-resistant materials Heat utilization	Bonus system for efficiency improvement

Technology	Countries	Efficiency	Autonomous improvement	Incentives
Power plant ORC	AT, DE	6–20% electrical	Autonomous improvement of this new technology	Bonus system for efficiency improvement Minimum efficiency standard
District heating	SE, FI, DK, EE, LV, AT	80–90% thermal	Improved heat recovery e.g. flue gas condensation; Boiler efficiency improvements	Minimum efficiency standard Investment subsidy
Manure (co–) digestion	DE, DK, AT, NL, IT, PT, HU	Not relevant (motor: 38– 42% electrical)	Heat utilization Avoid small-scale gas engines	Bonus system for efficiency improvement
Diesel engines on vegetable oils	IT, DE	40–48% electric	Heat utilization Avoid small-scale diesel engines	Bonus system for efficiency improvement Minimum efficiency standard
Lignocellulosic fuels for ground transport and aviation	USA, Brazil and European Union		Supply chain and biomass crop productivity + sugar-extraction efficiency from cellulosic material	Advanced biofuel quotas and incentives as per USA Renewable Fuels Standard (RFS 2) and European Union's Renewable Energy Directive

AT, Austria; DE, Germany; DK, Denmark; EE, Slovakia; FI, Finland; HU, Hungary; IT, Italy; SE, Sweden; LT, Lithuania; LU, Luxemburg; LV, Latvia; NL, the Netherlands; PT, Portugal.

Sources: modified from http://ec.europa.eu/energy/renewables/bioenergy/doc/2010_02_25_report_conversion_efficiency.pdf (Janssen et al., 2013).

in a combustion chamber. Biomass combustion facilities that generate electricity from steam-driven turbine generators have a conversion efficiency of 17–28%. Co-generation can increase this efficiency to almost 85%. Large-scale combustion systems use mostly low-quality fuels, while high-quality fuels are more frequently used in small application systems. It should be pointed out that in the last decade important advances in boilers of all types have been made.

The selection and design of any biomass combustion system is primarily determined by the characteristics of the fuel, availability of the feedstock, environmental constraints, cost of equipment and the size of plant.

There is increasing interest in wood-burning appliances for heating and cooking. Domestic wood-burning appliances include fireplaces, heat-storing stoves, pellet stoves and burners, central heating furnaces and boilers. There are various industrial combustion systems available, which – broadly speaking – can be defined as fixed-bed combustion (FxBC), fluidized bed combustion (FBC) and dust combustion (DC).

Co-firing

Co-firing is potentially a major option for the utilization of biomass, if some of the technical, social and supply problems can be overcome. Co-firing of biomass with fossil fuels, primarily coal or lignite, has received much attention, particularly in Europe and the USA, and important advances have been made. Biomass can be blended with coal in differing proportions; e.g. it is now considered that co-firing at rates of 20–50% biomass throughput (or more) is possible, a technique described as 'enhanced co-firing'. It is even anticipated that complete conversion of a coal boiler to biomass may be feasible. An alternative used often is pulverizing the coal and biomass simultaneously in existing coal mills, a technique usually termed 'co-milling' (or 'through-the-mill') that allows size reduction and drying of both the biomass and coal prior to the two fuels being burned together in the furnace (see Altawell, 2014).

The main advantages of co-firing include:

- the existence of an established market for CHP,
- relatively small investment required compared to a biomass-only plant (i.e. minor modification to existing coal-fired boilers),
- high flexibility in arranging and integrating the main components into existing plants (i.e. use of existing plant capacity and infrastructure),
- favourable environmental impacts compared to coal-only plants,
- potentially lower local feedstock costs (i.e. use of agro-forestry residues and energy crops; if present, productivity can be increased significantly),
- potential availability of large amounts of feedstock (biomass/waste) that can be used in co-firing applications, if supply logistics can be solved,
- higher efficiency for converting biomass to electricity compared to 100% wood-fired boilers; for example, biomass combustion efficiency to electricity would be close to 33–37% when fired with coal,
- planning consent is not required in most cases.

Gasification

Gasification is one of the most important research, development and demonstration areas in biomass for power generation, as it is the main alternative to direct combustion. Gasification is an endothermic conversion technology where a solid fuel is converted into a combustible gas. The importance of this technology lies in the fact that it can take advantage of advanced turbine designs and heat-recovery steam generators to achieve high-energy efficiency.

Gasification technology is not new; the process has been used for almost two centuries. For example, in the 1850s much of London was illuminated by 'town gas', produced from the gasification of coal. The main attractions of gasification are:

- higher electrical efficiency, e.g. 40% or more compared with 26–30% for combustion, while costs may be very similar,
- important developments on the horizon, such as advanced gas turbines and fuel cells,
- possible replacement of natural gas or diesel fuel used in industrial boilers and furnaces,
- distributed power generation where power demand is low,
- displacement of gasoline or diesel in an internal combustion (IC) engine.

There are many excellent reviews of gasification e.g. Kaltschmitt and Bridgwater (1997); Kaltschmitt et al. (1998); Walter et al. (2000).

Pyrolysis

The surge of interest in pyrolysis stems from the multiple products obtainable from this technology: for example, liquid fuels that can easily be stored and transported, and the large number of chemicals (i.e. adhesives, organic chemicals and flavouring) that offer good possibilities for increasing revenues. Any form of biomass can be used although cellulose gives the highest yields at around 85–90% wt on dry feed. Liquid oils obtained from pyrolysis have been tested for short periods on gas turbines and engines with some initial success, but long-term data is still lacking. There is ample literature for the interested reader to follow.

Combined heat and power (CHP)

CHP is a well-understood technology that is over a century old. Many manufacturing plants operated CHP systems in the late 19th century, although most abandoned the technology when utility monopolies began to emerge. Essentially, CHP is usually implemented by the addition of a heat exchanger that absorbs the exhaust heat, which is otherwise wasted, from an existing generator. The energy captured is then used to drive an electrical generator.

CHP is becoming fashionable primarily for the following reasons:

- energy efficiency: CHP is about 85% efficient, compared to 35–55% for most traditional electricity utilities;
- growing environmental concerns: it is estimated that each megawatt electrical of CHP saves approximately 1000 t of C/year;
- energy decentralization: recent world market projections indicate that the market for generators below 10 MW could represent a significant proportion of the large new capacity expected worldwide.

There is a large body of literature on CHP, for example see www.eeere.energy. gov of the US Department of Energy.

Advanced (second-generation) biofuels

Commercial demonstration of liquid biofuel production from lignocellulosic materials has only just begun (Janssen et al., 2013). Second-generation biofuels are based on feedstock comprising lignocellulosic biomass such as woody biomass, grasses, agricultural residues (bagasse, husks, shells, straw, stalks, leaves) and forestry residues (small roundwood, branches, leaves, sawdust, thinning, etc.). These feedstocks are available in abundance and can be harvested at a much lower cost than first-generation feedstocks. However, some of these feedstocks may have alternative uses. Some of the crops that could be grown for biofuel production include perennial grasses (switchgrass, napier grass and *Miscanthus*), short-rotation willows, hybrid poplar and eucalyptus, and micro-algae (Worldwatch Institute, 2006). Some feedstocks for biofuel production could become invasive species and a cautious approach is needed when introducing species into regions where they are currently not present (UNEP, 2009).

Conversion of cellulosic material to biofuels involves breaking down the biomass to release the sugars effectively locked in the complex lignocellulosic structures, followed by a range of conversion processes to convert the sugars to biofuels. Breaking down the lignocellulosic structure can be achieved through three different processes (House et al., 2012), as follows.

- Enzymatic processes, based on enzymes and microorganisms, are used to convert cellulosic and lignocellulosic components of the feedstock to sugars before their fermentation to produce ethanol or alternative fuel molecules. An indicative biofuel yield through the biochemical route (enzymatic hydrolysis of ethanol) is in the range of 110–300 l (average 200 l) per dry tonne of feedstock.
- Acidic processes use acid hydrolysis instead of, or in conjunction with, enzymes as described above.
- Thermochemical: in these thermal decomposition processes, pyrolysis/gasification technologies produce bio-oils or a synthesis gas (syngas; $CO + H_2$; the Fischer–Tropsch process) from which a wide range of

long-carbon-chain biofuels such as synthetic diesel or aviation fuel can be produced. An indicative yield of biodiesel through the processing of syngas in the Fischer–Tropsch reaction ranges from 75 to 200 l per dry tonne of feedstock, while syngas-to-ethanol ranges from 120 to 160 l per dry tonne. It is possible to produce a large range of fuels including drop-in biofuels from syngas by catalytically synthesizing the gas into longer-carbon-chain molecules that are similar or identical to those in petrol and diesel. This type of thermochemical process is called biomass-to-liquids (BTL). The derived fuels are often chemically and physically identical to their fossil analogues.

The future for bioenergy

Global changes in the energy market, particularly the emergence of shale gas, decentralization and privatization, have created new opportunities and challenges for both RE in general and biomass for energy in particular. Experiments in market-based support are changing the way we look at energy production and utilization.

It is notoriously difficult to forecast long-term energy demand. However, all indications point to a continuous growth in demand. Thus, the question is, how can this demand be met and what will be the most important resources? More specifically, what is likely to be the role of biomass for energy? Is biomass for energy finally reaching maturity?

Globally there is a growing confidence that RE in general is maturing rapidly in many areas of the world and not just in niche markets (see REN21, 2014 for a comprehensive review of the developing RE markets globally). It is important to recognize that the development of biomass energy will largely be dependent on the development of the oil and gas industry, and also other renewables, as it will be driven by similar energy, environmental, political, social, economic and technological considerations.

The 1970s were pioneering years providing a wealth of innovative ideas on RE that were further advanced in the 1980s when the computer revolution played a key enabling role. In the 1990s improvements in RE allowed the technology to meet emerging market opportunities, e.g. gasification, co-generation/CHP, etc. This opportunity was very much linked to the concern with climate change and the environment. The early 21st century may be dominated by a global policy drive to mitigate climate change, land competition, biodiversity, food prices and energy costs. It is essential that biomass energy is integrated with existing energy sources and thereby able to meet the challenges of integration with other RE and fossil fuels.

For bioenergy to have a long-term future it must be produced and used sustainably to demonstrate its environmental and social benefits in comparison to fossil fuels. The development of modern biomass energy systems is still at a relatively early stage, with most of the R&D focusing on the development of fuel supply and conversion routes that minimize environmental impacts. Far more attention needs to be paid the availability and affordability of bioenergy feedstocks and the intertwined factors with agriculture. Although the technologies are evolving quite rapidly, the R&D devoted to bioenergy is insignificant compared to that on fossil fuels, and this needs to be addressed. In addition, the development of biomass energy should be more closely integrated with other RE technologies, and with local capacity building, financing and the like.

Modernizing bioenergy will bring many benefits. Lugar and Woolsey (1999) wrote: 'Let us imagine, for example, that cellulose-based ethanol becomes a commercial reality. Imagine if hundreds of billions of dollars that currently flow into the coffers of a handful of nations, were to flow into millions of farmers, most countries would see substantial economic and environmental benefits as well as increased national security. With so many millions involved in production of ethanol fuel, it would be impossible to create a cartel. With new drilling oil technology, we will be able to make better use of existing resources and accelerate production, but would not be able to expand oil reserves.' 2014 saw a number of first-of-a-kind commercial-scale lignocellulosic ethanol plants come on stream, but enormous challenges remain in terms of commercial viability at scales that are sufficiently large to affect global energy markets. See for example http://energy.gov/eere/articles/four-cellulosic-ethanol-breakthroughs for the most recent developments in the USA.

The transportation system is more complex. IC engines and oil-derived fuels have dominated transportation for many decades and have been so successful that until recently prospects for radical alternatives were not taken seriously. Thus little research, development and demonstration has been directed to search for new alternatives.

It is only in the recent past that a combination of technological, environmental and socio-economic changes has forced the search for new alternatives that could challenge the dominance of the IC engine, and this is gaining momentum. However, which alternative(s) will prevail is still unclear, given the present stage of development and the range of alternatives under consideration from which no clear winners have emerged so far. In the short term the main challenge will be to find sound alternatives to fossil fuels that can be used in the IC engine; ethanol and biodiesel are currently in commercial use, while others such as hydrogen are emerging. In the longer term the challenge will be to find large-scale alternatives to fossil fuels that can be used in both existing IC engines and new propulsion systems. Two major challenges will be availability and sustainability of large-scale feedstocks and the technologies to process them. Biomass combustion (e.g. small and large boilers and other devices) are becoming a large-scale commercial reality.

Data on biomass production and use remains poor, despite considerable efforts to improve it. Consumption data often deals with the household sector, e.g. excluding data on many small enterprises. In particular, the modernization of biomass energy use requires a good information base. Only a handful of countries have a reasonably good database on biomass supply, although these are mostly based on commercial forestry data rather than bioenergy.

Despite increased recognition, biomass energy still does not receive the attention it deserves from policy makers, and it gets even less from educators. The following quotation is a good illustration (Openshaw, 2000):

> Wood energy, like the oldest profession, has been around since time immemorial; like prostitution, it is ignored or regarded as an embarrassment by many decision makers at the national and international level. However, for about half of the world's population it is a reality and will remain so for many decades to come.

Note

1 Another reason why bioenergy continues to grow, despite the fact that poor people (women and children) spend a considerable amount of time gathering firewood, dung, etc., is because biomass energy resources continue to be free in most cases. Free means that poor people do not pay cash; the time spent collecting wood, etc. is not taken into account or given economic value.

References and further reading

ALTAWELL, N. (2014) *The selection process of biomass materials for the production of bio-fuels and co-firing.* Wiley-IEEE Press, Chichester.

ANON (2001) *G8 renewable energy task force, final report.* IEA, Paris.

ANON (2003) *World ethanol production powering ahead.* F.O. Lichts, vol. 1(19), p. 139. www.agra-net.com.

ANON (2004) *World ethanol and biofuels report.* F.O. Lichts, vol. 7(3), pp. 129–135.

AZAR, C., LINDGREN, K. AND ANDERSON, B.A. (2003) Global energy scenarios meeting stringent CO_2 constraints: cost-effective fuel choices in the transportation sector. *Energy Policy* 31, 961–976.

BHATTACHARYA, S.C. (2004) Fuel for thought: the status of biomass energy in developing countries. *Renewable Energy World* 7(6), 122–130.

BHATTACHARYA, S.C., ATTALAGE, R.A., AUGUSTUS LEON, M., AMUR, G.Q., SALAM, P.A. AND THANAWAT, C. (1999) Potential of biomass fuel conservation in selected Asian countries. *Energy Conversion and Management* 40, 1141–1162.

CHUM, H., FAAIJ, A., MOREIRA, J., BERNDES, G., DHAMIJA, P., DONG, H., GABRIELLE, B., GOSS ENG, A., LUCHT, W., MAPAKO, M., MASERA CERUTTI, O. ET AL. (2011) Bioenergy. In *IPCC special report on renewable energy sources and climate change mitigation*, O. Edenhofer, R. Pichs-Madruga, Y. Sokona, K. Seyboth, P. Matschoss, S. Kadner, T. Zwickel, P. Eickemeier, G. Hansen, S. Schlömer and C. von Stechow (eds). IPCC, Cambridge.

CREUTZIG, F., RAVINDRANATH, N.H., BERNDES, G., BOLWIG, S., BRIGHT, R., CHERUBINI, F., CHUM, H., CORBERA, E., DELUCCHI, M., FAAIJ, A. ET AL. (2014) Bioenergy and climate change mitigation: an assessment. *Global Change Biology Bioenergy* DOI: 10.1111/gcbb.12205.

DALE, B.E., ANDERSON, J.E., BROWN, R.C., CSONKA, S., DALE, V.H., HERWICK, G., JACKSON, R.D., JORDAN, N., KAFFKA, S., KLINE, K.L. ET AL. (2014) Take a closer look: biofuels can support environmental, economic and social goals. *Environmental Science Technology* 48, 7200–7203.

FAO/WE (2003) *Bioenergy terminology*. FAO Forestry Department, FAO, Rome.

GEA (2012) *Global energy assessment: toward a sustainable future*. Cambridge University Press, Cambridge and International Institute for Applied Systems Analysis, Laxenburg, Austria.

HALL, D.O. AND OVEREND, R.P. (1987) Biomass forever. In *Biomass: regenerable energy*, D.O. Hall and R.P. Overend (eds). John Wiley & Sons, Chichester, pp. 469–473.

HALL, D.O. AND RAO, K.K. (1999) *Photosynthesis*, 6th edition. Cambridge University Press, Cambridge.

HALL, D.O., HOUSE, J.I. AND SCRASE, I. (2000) Overview of biomass energy. In *Industrial uses of biomass energy: the example of Brazil*, F. Rosillo-Calle, S. Bajay and H. Rothman (eds). Taylor & Francis, London, pp. 1–26.

HALL, D.O, ROSILLO-CALLE, F. AND WOODS, J. (1994) Biomass utilisation in households and industry: energy use and development. *Chemosphere* 29(5), 1099–1119.

HOOGWIJK, M., DEN BROEK, R., BERNDES, G. AND FAAIJ, A. (2001) A review of assessments on the future of global contribution of biomass energy. In *1st World Conf. on Biomass Energy and Industry*, Seville. James and James, London, vol. II, pp. 296–299.

HOUSE, J.I., BELLARBY, J., BÖTTCHER, H., BRANDER, M., KALAS, N., SMITH, P., TIPPER, R. AND WOODS, J. (2012) Mitigating climate risks by managing the biosphere. In *Understanding the earth system*, S.E. Cornell, C.I. Prentice, J.I. House and C.J. Downy (eds). Global Change Science for Application. Cambridge University Press, Cambridge, Chapter 7.

IEA (2002a) *Energy outlook 2000–2030*. IEA, Paris.

IEA (2002b) *Handbook of biomass combustion and co-firing*. International Energy Agency (IEA), Task 32. IEA, Paris.

IEA (2010) *Energy technology perspectives 2010: scenarios and strategies to 2050*. OECD/IEA, IEA, Paris.

IEA (2011a) *Technology roadmap: biofuels for transport*. OECD/IEA, IEA, Paris.

IEA (2011b) *World energy outlook 2011*. OECD/IEA, IEA, Paris.

IEA (2012a) *Energy technology perspectives*. OECD/IEA, 2012. OECD/IEA, IEA, Paris.

IEA (2012b) *Technology roadmap: bioenergy for heat and power*. OECD/IEA, IEA, Paris.

IPCC (2014) *Bioenergy: climate effects, mitigation options, potential and sustainability implications*. Appendix to Chapter 11 (AFOLU) Final Draft. IPCC WGIII AR5. IPCC, *JEM* Geneva.

IPCC-TAR (2001) Climate change 2001: mitigation. In *Third assessment of the IPCC*, B. Mentz et al. (eds). Cambridge University Press, Cambridge.

JANSSEN, R., TURHOLLOW, A.F., RUTZ, D. AND MERGNER, R. (2013) Production facilities for second generation biofuels in the USA and the EU: current status and future perspectives. *Biofuels, Bioproducts and Biorefining* 7, 647–665.

KALTSCHMITT, M. AND BRIDGWATER, A.V. (eds) (1997) *Biomass gasification and pyrolysis: state of the art and future prospects*. CPL Press, Newbury.

KALTSCHMITT, M., ROSCH, C. AND DINKELBACH, L. (eds) (1998) *Biomass gasification in Europe*. EC Science Research & Development, EUR 18224 EN, Brussels.

KARTHA, S., LEACH, G. AND RJAN, S.C. (2005) *Advancing bioenergy for sustainable development: guidelines for policymakers and investors*. Energy Sector Management Assistance Programme (ESMAP) Report 300/05. The World Bank, Washington, DC.

LUGAR, R.G. AND WOOLSEY, J. (1999) The new petroleum. *Foreign Affairs* 78(1), 88–102.

NIGHT, B. AND WESTWOOD, A. (2005) Global growth: the world biomass market. *Renewable Energy News* 8(1), 118–127.

OPENSHAW, K. (2000) Wood energy education: an eclectic viewpoint. *Wood Energy News* 16(1), 18–20.

PERLIN, J. AND JORDAN, P. (1983) Running out: 4200 years of wood shortages. *The Convolution Quarterly*, Spring.

REN21 (2014) *Renewables 2014 global status report*. Renewable Energy Policy Network for the 21st Century (REN21). REN21 Secretariat, Paris.

ROSILLO-CALLE, F. (2003) *Public dimension of renewable energy promotion: sitting controversy in biomass-to-energy development in the UK*. EPSRC Internal Report, Kings College London, London.

ROSILLO-CALLE, F. (2006) Biomass energy. In *Landolf-Bornstein handbook, vol 3: energy technologies*. Springer, Germany, pp. 334–413.

ROSILLO-CALLE, F. AND HALL, D.O. (1992) Biomass energy, forests and global warming. *Energy Policy* 20, 124–136.

ROSILLO-CALLE, F. AND WALTER, A.S. (2006) Global market for bioethanol: historical trends and future prospects. *Energy for Sustainable Development*, Special Issue 10(1), 20–32.

SCHUBERT, H.R. (1957) *History of the British iron and steel industry*. Routledge & Keagan Paul, London.

SIMS, R.E.H. (2002) *The brilliance of bioenergy: in business and in practice*. James and James, London.

SLADE, R., BAUEN, A. AND GROSS, R. (2014) Global bioenergy resources. *Nature Climate Change* 4, 99–105.

UNEP (2009) *Towards sustainable production and use of resources: assessing biofuels*. www.unep. fr/energy/bioenergy.

WALTER, A., BAIN, L., OVEREND, R.P., CRAIG, K.R., FAAIJ, A., BAUEN, A., MOREIRA, J.R., BEZZON, G. AND ROCHA, J.D. (2000) New technologies for modern biomass energy carriers. In *Industrial uses of biomass energy: the example of Brazil*, F. Rosillo-Calle, S. Bajay and H. Rothman (eds). Taylor & Francis, London, pp. 200–253.

WOODS, J. AND HALL, D.O. (1994). *Bioenergy for development: environmental and technical dimensions*. FAO, Rome.

WOODS, J., LYND, L.R., LASER, M., BATISTELLA, M., DE CASTRO VICTORIA, D., KLINE, K. AND FAAIJ, A.P.C. (2015) Land and bioenergy. In *Bioenergy and sustainability*, G.M. Souza, R. Victoria, C. Joly and L. Verdade (eds). SCOPE (Scientific Committee on Problems of the Environment), BIOEN, BIOTA and PFPMCG. www. bioenfapesp.org/scopebioenergy (in press).

WORLDWATCH INSTITUTE (2006) *Biofuels for transportation: global potential and implications for sustainable agriculture and energy in the 21st Century – extended summary*. German Federal Ministry of Food Agriculture and Consumer Protection, Washington DC.

ZERVOS, A., LINS, C. AND SCHAFER, O. (2004) Tomorrow's world: 50% renewables scenarios for 2040. *Renewable Energy World*, 7(4), 238–245; www.erec-renewables.org.

2 General introduction to the basis of biomass assessment methodology

Frank Rosillo-Calle, Peter de Groot,
Jeremy Woods, Vineet V. Chandra and
Sarah L. Hemstock

Introduction

This chapter introduces some of the main problems in measuring the use and supply of biomass, outlines the system for classifying biomass used in this handbook and sketches general methods for assessing biomass. It also looks, albeit briefly, at biomass flow charts, units for measuring biomass (e.g. stocking, moisture content and heating values), weight versus volume, calculation of energy values, implications for setting up a biomass energy plant and, finally, possible future trends. Most of the specific issues covered here are complemented in much greater detail in Chapter 3. In short, this chapter summarizes in some way all the major issues presented in this handbook.

Despite the importance of biomass for energy, there is surprisingly little reliable and detailed information on the consumption and supply and no standardized system for measurements and accounting procedures, in particular in the case of traditional applications. This serious lack of information is preventing policy makers and planners from formulating satisfactory sustainable energy policies.

This handbook gives methods for obtaining estimates of aggregated data for biomass for energy at the country and regional levels down to detailed disaggregated data at the local level. The emphasis is on traditional bioenergy applications, although modern uses of biomass for energy are also considered. None of these methods is perfect: each has associated penalties in time, human resources and money, and each will provide differing types of information with varying degrees of accuracy. However, used appropriately these methodologies will pinpoint critical problems in biomass production and supply. As bioenergy is very site-specific you will need to use your own judgement according to the circumstances as to the most appropriate way to carry out your assessment.

Problems in measuring biomass

There are many problems associated with measuring biomass, but they generally fall into three groups, as follows.

1   *The difficulty in physically measuring the biomass.* The main problems here arise in distinguishing between the potential and actual supply, and in measuring variability. Biomass is often collected over a wide area from a range of vegetation types. Estimating the supply of biomass that is theoretically available for energy can be difficult, particularly if detailed data is needed. But when an accurate estimate of potential supply is obtained, the actual supply will then depend upon the access to this biomass. Topography will dictate how difficult the biomass is to collect, while local laws, traditions or customs may also restrict access to certain areas for the collection of biomass. Thus accessibility rather than total biomass available may be the most important parameter.

2   *The multiple and sometimes sequential uses to which biomass is put.* An assessment of the consumption and supply of biomass is very different from a similar assessment for a commercial fuel such as kerosene. While kerosene is used as a fuel for heating and lighting, biomass provides a range of essential and interrelated needs, especially in developing countries. These benefits include not only energy but also food, fodder, building materials, fencing, medicines and more. Biomass is rarely, if ever, planted specifically for fuel: wood that is burned is often what is left over from some other process. Biomass energy should therefore always be looked at in the context of the other benefits that biomass provides, and never just from the point of view of a single sector.

Biomass products can be multi-purpose, e g. sugar cane bagasse is used as fuel, as animal feed after hydrolysis, in the construction industry and for making paper. Other forms of biomass are modified from one energy form into another, e.g. wood to charcoal, dung to biogas and fertilizer, and sugar to ethanol. Thus, it may be important to measure processed biomass as an actual or a potential energy source.

It is also necessary to know the quantity of, say, woody raw material available in order to estimate the quantity used to produce charcoal, pellets, etc. For example, if rice husks have a use as a boiler and kiln fuel, the annual production rate at specific sites is required to assess their economic use and physical availability for processing into, say, briquettes or pellets.

3   *The many different units that are used to measure biomass.* For example, firewood is frequently measured by weight, sometimes by volume and often by the bundle or backload. Conversion between these different types of measurement can be tricky: the weight of a volume of wood will depend on its moisture content, and while a backload is a convenient measure for the woman collecting the firewood, a surveyor will have to take an average of several backloads to get a quantitative measurement.

The next section lists some general points for consideration before embarking on a biomass assessment.

The ten commandments of biomass assessment

Below there are what we call the 'ten possible commandments' that may help you to avoid the worst pitfalls in assessing biomass.

1 *Never confuse consumption with need.* Need may exceed consumption if biomass is in short supply or expensive. Consumption is largely dependent on the perceived costs of a biomass resource. Costs reflect both supply and accessibility. Try to estimate 'basic energy needs', defined as the minimum energy requirement for basic activities such as cooking, heating and lighting. A developmental component may be added to allow for household and small scale cottage industrial activities.

2 *Do not separate consumption from supply.* For convenience, the supply and consumption of biomass are treated separately in this handbook. However, data on the availability and on the use of biomass energy is often collected at the same time, even though different people may be responsible for collecting the two sets of data.

3 *Make your assumptions explicit.* Most empirical methods will entail implicit or explicit assumptions, or even be influenced by decisions made on the nature or aims of the project. Any underlying motives should be clearly stated to allow for consistent accounting and to make it possible to compare results over time.

4 *Your data requirements must be driven by your problem.* Data collection it is not an end in itself. Estimates of consumption are made to provide a basis for action. Collecting comprehensive, disaggregated data will place an enormous burden on your resources. You have to decide how detailed the data should be, and when aggregated information (or extrapolation) will be sufficient.

5 *Be aware of the danger of devoting too many resources to data collection and too few to analysis.* Analysis of the data takes time, and must be carefully planned. Ask yourself if you have the necessary resources to carry out a highly detailed analysis. If not, plan accordingly and explain the reasons why.

6 *Do not ignore it because you cannot measure it.* It is not always possible to quantify the demand for biomass. It is therefore advisable to incorporate a broad base of empirical information into the regular project reviews. Talk to the local population; involve locals as much as possible.

7 *Do not be beguiled by averages.* Demand figures given as averages should clearly be understood as central tendencies within a distribution. If sample sizes are small, average figures are of little value. Averages and statistical information are valuable parts of the data presentation, but always treat average figures with caution.

There is often considerable variation in energy consumption patterns, not just between countries but also between areas separated only by a few kilometres, between and within ecological zones, and over time. These problems are further compounded where data is collected according to administrative boundaries. These rarely, if ever, match ecological zones.

Confusion often arises when a comparison is attempted between data collected according to political and ecological zones.

8   *There is not one single simple solution, so distrust the simple single answer.* Energy is only one of the many uses (and may not be the main use) to which biomass is put. The estimation of the supply and consumption of bioenergy is therefore not a simple matter, and there is no single method for drawing up an accounting system.
9   *The users are the best judges of what is good for them.* After all, a major objective of a biomass assessment is to help the consumer. A good understanding of the socio-economics, cultural practices and the needs of the community are therefore essential.
10 *Be flexible, and modify any of the above if it seems sensible to do so.* Remaining flexible may prove to be the most important rule of all.

All these issues are dealt in greater detail in the following pages.

Some other general considerations when making a biomass assessment

The methods employed to assess biomass resources will vary, depending on:

*   the *purpose* for which the data is intended,
*   the *detail* required,
*   the *information already available* for the particular country, region or local site.

The following are a number of steps that will help you with your assessment.

*   *Clearly define the purpose and objectives* of your assessment. Why is it needed? What does it seek to achieve?
*   *What is the audience that you are addressing?* For example, policy makers, planners and project managers require information expressed in different ways than the traditional user.
*   *Decide on the level of detail required for your data.* Policy makers will require aggregated information, for example, while project personnel will probably require highly detailed, disaggregated data. Where fine detail is required for the implementation of projects, exhaustive investigation is necessary to provide a clear, unambiguous report on each type of biomass resource and to provide a detailed analysis of availability, accessibility, convertibility, present use pattern and future trends.
*   *Decide upon a system of biomass classification.* This handbook divides biomass into eight major categories, which will allow you to use similar methods of assessment and measurement for each type of biomass (see below).
*   *Identify critical areas in biomass supply and demand.* This is essential if policy makers and planners are to make the right decisions.

- *How much data is already available?* A thorough literature search involving the co-ordination, assembly and interpretation of information from existing sources, including national, regional and local databases, and statistics produced by both government and non-government organizations, may save a great deal of time, and avoid unnecessary duplication of effort. A lot of data may already be available in the form of maps and reports, for example.

Care is needed when using data from existing sources. The Food and Agriculture Organization (FAO) is a major source of data on the supply of biomass, particularly woody biomass data, published in, for example, the *World Forest Inventories*. The FAO and other agencies usually derive their information from country reports, many of which have not been updated for a considerable time, and which may not reflect a true picture for a variety of reasons (e.g. lack of resources, or a bias against bioenergy, particularly in traditional applications). In addition, statistics are for the most part concerned only with commercial applications (e.g. in the case of forests, with the stem volume) compiled primarily from the assessment of industrial wood. In many developing countries information on growing stock and yield tends to be incomplete and is generally considered inaccurate.

Most published data on wood biomass does not consider trees outside forests or woodlands, and thus ignores the fact that much collection of fuelwood takes place outside the forest. Important sources of fuelwood such as small-diameter trees, shrubs and scrub and branch wood are often ignored in these studies. Fuelwood is normally an integral part of the informal economy and rarely enters into the official statistics.

The lack of standard methods or units for documenting the supply or consumption of biomass may make it difficult to compare or incorporate data from previous surveys. Convert existing data to standard units to allow easy comparisons between localities whenever possible.

- *Decide how to measure biomass, and which units to use.* Rural women know well how much fuelwood in its different forms is needed to cook a meal. However, you need to obtain more than empirical knowledge using accurate scientific principles. Unfortunately there is no standard method for measuring biomass used for fuel. In future, the need for standardized methods for measuring biofuels will increase as new industrial applications using biomass come on stream, as these plants will need detailed and compatible information on the type and quantity of raw material they require. This is already happening with modern applications.

There are various methods and techniques for measuring biomass, by volume, weight or even length (see below). For some species, particularly those used commercially, techniques for assessing availability and potential supply are readily available. The commercial forestry sector traditionally measures biomass, especially woody biomass, by volume. However, biomass fuels are usually irregularly shaped, e.g. small branches, twigs, split wood and stalks, for which volume is an awkward method of measurement. Thus for biomass energy the most appropriate measuring method is weight rather than volume.

For non–commercial species, and locations where a multitude of differing trees, shrubs, etc., are present, it is likely that no assessment methodologies exist. You may be able to adapt some methods used in assessments carried out in other locations, or even from the commercial forestry sector. However, you need to keep in mind that the supply and end use may be very different in other locations, and that the methods and techniques used in the commercial sector may be unsuitable.

- *Consider demand and supply analysis.* Where data on biomass supply is initially too difficult to obtain, the use of demand analysis data may be useful to fill the information gaps.
- *Differentiate between potential and actual supply of biomass.* Obtaining an accurate estimate of the potential supply can be a problem in itself since it depends on many factors such as topography, local laws, traditions, ownership rights, etc.
- *Aim to collect time-series data.* Only data collected over a number of seasons, say 5 years, will show trends in use, and allow for climatic variation (both annual and seasonal).
- *Monitor the results of any programme.* This will provide essential feedback to confirm that the biomass programme is meeting energy demands, and that it is sustainable.
- *Proper measurement of the heating value.* The energy content of biomass varies according to its moisture and ash content, both of which must be taken into account.
- *Sound decisions require sound, accurate data.* Most published data on biomass energy usually considers only recorded fuelwood removal from forests, whereas biofuels harvested from agricultural land are largely ignored. As a result, this ignores large areas of the actual fuel supply collected outside forests, such as twigs and small branches, shrubs, etc. As said already, traditional applications are often an integral part of the informal economy and hardly enter official statistics.

The variability of biomass and the skills required to make the most efficient use of available biomass adds a further complication to estimating effective end use. However, this information is only vaguely quantified, and rarely recorded. As a result, the figures for biomass consumption are rarely accurate (see Chapter 5).

Biomass

Processed biomass

In addition to measuring the potential supply of biomass such as growing stock, annual yield of woody and non-woody biomass, annual crop production, animal residues, etc., it is important to measure the supply of processed biomass such as sawmill waste and charcoal, as this is often an important source of energy. Estimates of processed biomass should probably not pose any serious problem since a lot of information from similar commercial activities can be obtained elsewhere. For example, dedicated energy plantations in Brazil use basically the same biomass measurement techniques that are applied to other commercial plantations (see also Chapter 4).

Non-woody biomass

Non-woody biomass, particularly agricultural residues, animal waste and herbaceous crops, is a major source of energy. However, the use of non-woody biomass can be quite localized. For example, the large-scale use of dung is largely confined to a few countries such as India. The methodology used for estimating non-woody biomass depends on the type of material and the quantity of statistical data available or deducible. Accurate data – on national, regional or local levels – is required. For agricultural residues, for example, you should only be concerned with gathering data about what is being used as a fuel, not with the total non-woody biomass production on a given site. The same applies to animal wastes. Measuring supply and consumption of dung will not be of any value unless dung is an important source of energy (see Appendix 2.1).

Secondary fuels

Secondary fuels obtained from raw biomass (producer gas, ethanol, methanol and briquettes, etc.) are increasingly being used in modern industrial applications. New methods are being developed to deal with the measurement of industrial uses of bioenergy (see below and Chapter 4).

Changing fuel consumption

Changing fuel-consumption patterns are important indicators of socio-economic and cultural changes. Thus, for an accurate estimate of biomass consumption, it is important to capture these behavioural changes. Asking the right questions to the right people is a key to getting accurate data.

Modelling

A model may be defined as 'a simplified or idealized representation of a system, situation or process, often in mathematical terms devised to facilitate calculations and predictions'. Models serve as a learning aid and a tool to analyse possible interactions in the system under study. Models do not provide accurate forecasts, particularly in extensive and diverse rural energy situations. Thus the limited application of modelling to rural energy has resulted in models failing to play the kind of prominent role that was envisaged a decade or two ago. However, if you decide to use a model (and modelling has improved significantly), ensure that the parameters are clearly defined so that only key variables are selected for observation or measurement. This avoids the collection of unnecessary data. For woody biomass, it is better to define the model in mathematical terms of stand growth or yield, for example, as this will make data requirements easier to identify and enable more precise estimates from your model. The choice of modelling technique depends upon the data available and the use to which your estimates will be put. There are different types of model, which can be classified according several criteria, as follows.

- *Application.* For example, is the information resulting from your model to be used for planning or policy or at the regional, national or village level?
- *Objectives.* Ask the question, will the outputs from your model be used for demand forecasting, resource assessment, least-cost supply planning, investment appraisal, economic development, environmental assessment, integrated planning, etc?
- *Style.* In what manner will the data be used: dedicated or generic, degree of flexibility, the level of integration, specific nature, use of scenarios?
- *Technique.* For example, which variables and interactions are endogenous, do you wish to include optimization methods, should the data be dynamic or static, are you interested in demand- or supply-led models, which financial tools will you apply, do you want to carry out an environmental impact assessment? (Smith, 1991).

Databases

Biomass data comprises many components in its type, origin, production, availability, conversion and end use, and each of these components may be further divided into many sub-components. The availability of a good databank can facilitate the capture of this information in a way that it can be usefully retrieved in an easily understandable and meaningful manner to allow the policy maker and energy planner to make informed decisions. Fortunately, good new databases have been set up in recent years that, while still having a long way to go, represent a major improvement on past efforts. The internet and the increasing number of bioenergy-related networks have also facilitated significantly the task of compiling data. But extreme care should be taken which websites are visited: use official and well-recognized ones, as there is considerable amount of inaccurate data that often reflects particular interests (see also Chapter 1).

Remote sensing

Remote sensing is an analysis of detailed land use patterns based on aerial photography or high-resolution satellite imagery (in particular SPOT and Thematic Mapper). Satellite photography can be used to determine areas of dense woody biomass, but it cannot be used to give information on growing stock or annual increments. Aerial photography is better for fairly dense woodlands, as it gives a higher definition than satellite imagery; it also may allow the measurement of height, crown and even the stem diameter of scattered trees. If woodlands are relatively undisturbed, and are representative of all or most age and diameter classes, then approximate volumes and weights can be obtained from crown cover. For farm trees there is no substitute for field surveys, as these trees are intensively managed. To emphasize the point, remote sensing can form the basis for a comprehensive land use analysis. However, remote sensing data must be verified by detailed fieldwork, which can be costly. Remote sensing techniques have a lot to offer and are described in detail in Chapter 8, which focuses on agricultural land use.

Land use assessment

Data concerning local land use patterns will be required for project implementation, when an up-to-date analysis is essential. Any information, whether already existing or based on new field surveys or remote sensing, should be carefully verified in the field. It is advisable to begin collecting data at the most general, aggregated level. More detailed information can be collected later, if necessary. Land use assessment is important since it can be a key factor in determining actual biomass accessibility: one of the hardest and most essential tasks of any study of biomass. Also, land use change has become a very sensitive area when it concerns the production of energy (see Chapter 7).

The primary objective of a land use assessment, or a land evaluation, is to improve the sustainable management of land resources. A land evaluation analyses data about the soils, climate, vegetation, etc., and focuses upon the properties of the land, its functions and its potential. Land evaluations may be used for many purposes, ranging from land use planning to explore the potential for specific land uses or the need for improved land management or for the control of land degradation.

Most current rural development aims to alleviate economic and social problems, in particular hunger and poverty. Land evaluation is a useful tool as there is a clear focus upon the people, the farmers, the rural communities and other stakeholders in the use of land resources.

There is now a growing need for land evaluation, particularly wherever the problems of farmers are caused or compounded by problems of the land, e.g. decline in soil fertility, erosion and increased frequency of droughts due to climatic change. However, an objective and systematic assessment of the suitability of land resources for diverse uses becomes all the more necessary as growing populations create conflicting demands for uses of land other than agriculture, such as urbanization, transport, recreation and nature reserves, fuels and products.

Land suitability evaluation, the methodology set out in the Framework for Land Evaluation and later expanded in the Agro-ecological Zones methodology, was conceived and applied primarily in terms of sustainable production of crops and pastures, and in forestry. However, following the broader definitions of land and land resources, there is a growing need to address issues related to the capacity of the land to perform multiple economic, social and environmental functions (see Chapter 7).

Land performs a number of key interdependent environmental, economic, social and cultural functions essential for life. The land can only provide these services if it is used and managed sustainably. The land also provides services that are useful to humans and other species (e.g. water supply, carbon sequestration).

You may need to address the still widely perceived view that biomass energy competes directly with food production. In most cases food and fuel are complementary to each other, but if there is a conflict with food production you should identify it (see Case study 9.4).

Changes in land use patterns over time can provide a useful understanding of the evolution of local biomass resources, and enable predictions of likely resource availability in the future. These changes can be measured in various complementary ways:

- by the analysis of historical remotely sensed data, e.g. aerial photography, if this is available,
- with the use of official agricultural data,
- through detailed discussion in the field with the local population.

Steps involved in a land use assessment

Assessment of land suitability is carried out by a combination of matching constraints with crop requirements, and by modelling of potential biomass production and yield under constraint-free conditions. This activity is normally carried out in two main stages, in which first the agro-climatic suitability is assessed, and secondly the suitability classes are adjusted according to edaphic or soil constraints. Each stage comprises a number of steps, which are listed as follows (Figure 2.1).

- Stage 1: Agro-climatic suitability and agronomical attainable yields.
    1 Matching the attributes of temperature regimes to crop requirements for photosynthesis and phenology as reflected by the crop groups, determining which crops qualify for further consideration in the evaluation.
    2 Computation of constraint-free yields of all the qualifying crops taking into account the prevailing temperature and radiation regimes in each length-of-growing-period zone.
    3 Computation of agronomical attainable yields by estimating yield reductions due to agro-climatic constraints of moisture stress, pests and diseases, and workability for each crop in each length of growing period zone.

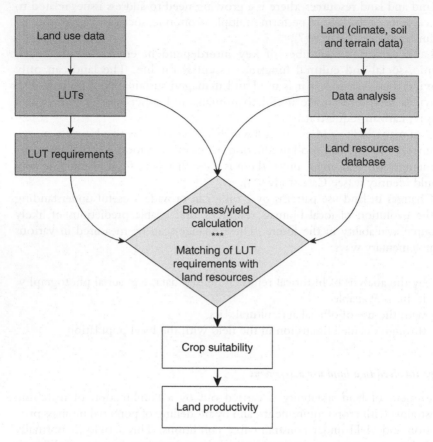

Figure 2.1 Steps involved in land use assessment. LUT, land utilization type.

- Stage 2: Assessment of agro–edaphic suitability based on soil constraints.

  4 Comparison of the soil requirements of crops with the soil conditions of the soil units described in the soil inventory, at different levels of inputs.

  5 Modifications are inferred by limitations imposed by slope, soil texture and soil phase conditions.

Apart from step 2, which involves a mechanistic model of biomass production and crop yield, all the above procedures involve the application of rules which are based on the underlying assumptions which relate land suitability classes to each other, and to estimates of potential yields under different input levels. Many of these rules were derived from expert knowledge available when the first FAO Agro-ecological Zone (or AEZ) study was undertaken in late 1970s see e.g. Fischer et al. (2006) and they should be regarded as flexible rather than rigid. The number of suitability classes, the definition of management and input levels, and the relationships between them can be modified according to increasing availability of information and the scope and objectives of each particular AEZ investigation. Results on a worldwide basis for more than 150 crop varieties were published by FAO/IIASA in 2000 and 2002. The results are available at http://webarchive.iiasa.ac.at/Research/LUC/SAEZ/pdf/gaez2002.pdf.

Changing land use

It is extremely valuable to examine the change in land use over time. This can be done in two complementary ways: (1) by analysis of historical remotely sensed data if this is available and (2) by detailed discussion in the field with the local population. Changes in land use patterns can be related to the issues under examination to help understand the evolution of local biomass resources.

The method of assessing land use at the planning level depends entirely upon the circumstances. The national policy-level analysis should form the basis for this assessment, but further enhancement and disaggregation will almost certainly be needed. A fully disaggregated land use assessment is not normally appropriate, but a proper analysis of agro-climatic zones will be necessary. If this information is not available, some form of remote sensing is the most viable option. The particular technique chosen will depend on local circumstances.

To sum up, unless detailed studies from the area in question already exist, some detailed local field work at chosen sample sites is desirable. These surveys should be minimized so that they are not too costly or demanding on time. However, they should be detailed enough to provide a true picture.

Remember, a major factor that will determine the nature of your assessment will be financial and human resources available to you. Almost inevitably, you will have to compromise between the kind of assessment that is desirable and the assessment that is possible given the resources at your disposal.

This being the case, it is important to use your resources as effectively as possible. For example, the time you devote to the assessment of a particular biomass resource should be determined by its relative importance to consumers. Unproductive woodland does not deserve the same degree of attention as woodland from which a large amount of firewood is collected. Keep in mind consumer preferences and other available alternatives.

The importance of field surveys

Field surveys are frequently used to generate much needed data, but they are not without pitfalls. On-the-ground surveys are complex, time-consuming and expensive. Therefore, you should consider a field survey only if other approaches prove insufficient. You should also consider how to supplement structured field surveys with, say, informal talks with the local population, as this is often the best way to collect data and to gain their confidence, and to get a degree of understanding of local cultural, social and economic practices.

It is important that you take care to design your survey so that you ask the right questions, and to employ competent people for its implementation. Focus initially on those issues you consider absolutely necessary. In most cases, surveys should be carried out by a small multi-disciplinary team, with the right skills and a well-structured questionnaire to ensure all relevant information is entered accurately.

Deciding about surveys in general

Before undertaking the survey, it is necessary to consider seven basic questions.

What is the problem?

A biomass assessment is carried out to implement some kind of action. The nature of the action required will determine the nature of the assessment. It is therefore essential to define the objectives of the assessment and the audience to be addressed. The nature of the problem tends to change over time and according to the nature of the biomass. For example, the problems posed by a lack of firewood for cooking are very different from those caused by a lack of biofuels for modern applications, such as co-generation.

Who is your audience?

It is convenient to think of various types of audience, for example policy makers, energy planners and project implementers, each of whom will have different objectives and perceptions about biomass. The energy planner would have a good understanding of energy supply and demand, although if they come from the conventional energy sector they could be biased. Understanding your audience is therefore very important in order to assist them to plan and implement biomass energy interventions.

How detailed should the information produced be?

The level of information required can only be decided once you define the reasons for the survey, and the professionals for whom the results will be addressed. The type of information, and the manner in which it is presented, ranges from a generalized but succinctly presented picture for the policy maker to detailed scientific data for the project officer. Your assessment may contain any level of precision of detail between these two extremes, or it may contain sections addressed to different audiences. Whatever the case, the method you adopt to collect and present your data is determined in the first instance by the information needs of the people you wish to address. Your data must be a reflection of the problem you are addressing; what changes is the way you present it so people can understand you. But, in addition, be honest and do not hide reality because your audience may not like your data.

What resources are available for the survey?

The financial and human resources available are the next important factors that will determine the nature of your assessment, as indicated. You will almost certainly have to compromise between the kind of assessment that is desirable and what is possible given the resources at your disposal. Work within your means. Explain the situation so that your audience understands your limitations.

Use your resources as effectively as possible. The time devoted to the assessment of a particular biomass resource should reflect its relative importance to consumers. Unproductive woodland does not deserve the same degree of attention as woodland from which a large amount of, say, firewood is collected.

Is a field survey necessary?

Field surveys can produce accurate and detailed data. However, they are often complex, time-consuming, expensive and demanding of skilled personnel, especially in remote areas and harsh terrains. Explore alternatives such as analysis of existing data or collaboration with national surveys. Only consider mounting a field survey if there is no alternative, or if the areas involved are not too large.

What are the scope and quality of existing data?

Much data is often already available, for example in the form of maps and reports. Make a thorough search of the available literature before you begin your survey. In recent years a lot of data has been collected by many international and government agencies. Another good source is the web, but as indicated above you need to be careful which source you are using: the web, with all its benefits, is also populated by unreliable and conflicting data. Information on the availability of woody biomass is often patchy, especially in developing countries that do not have the resources to carry out large detailed surveys. In particular, there is a pressing need to improve the statistics on fuelwood, which constitutes the bulk of the wood used in many developing countries, when traditional uses are concerned. Most published data on wood biomass considers only recorded (official) fuelwood removal from forest reserves (though it is often not made clear that this is the case). Do not forget that traditional uses of biomass are part of the informal economy, that therefore official consumption figures do not reflect the true picture. The official figures also ignore large areas of actual fuel supply, including:

- unrecorded removals from forests,
- trees on roadsides and community and farm lands,
- small-diameter trees,
- shrubs and scrub in forests,
- branch wood.

What equipment is needed?

Having decided on the nature and detail of the resource survey, the next step is to make an accurate assessment of a country, region or locality's natural resources. These should include:

- land type,
- vegetation type,
- soil composition,
- water availability,
- weather patterns.

Biomass classification

The system adopted in this handbook divides biomass types into eight categories. This has the advantage that similar methods of assessment and measurement for each type allow comparisons to be made between biomass types in different locations. You may wish to use a more refined classification system, but whatever method you adopt make sure it is clearly specified.

1   *Natural forests/woodlands.* These include all biomass in high-standing, closed natural forests and woodlands. Forests are defined as having a canopy closure of 80% and more, while woodland has a canopy closure of between 10 and 80%. This category will also include forest residues.

2   *Forest plantations.* These plantations include both commercial plantations (for pulp, paper and furniture) and energy plantations (trees dedicated to produce energy such as charcoal, and other energy uses). The total contribution of bioenergy in the future will partly depend on the potential of 'energy forestry/crops plantations' since the potential of residues is more limited. In the 1970s, 1980s and 1990s energy plantations were heralded as the major source of biomass energy in the future, but more recently this potential has been considered to be far more limited given the multiple demands on land use and agriculture.

3   *Agro-industrial plantations.* These are forest plantations specifically designed to produce agro-industrial raw materials, with wood collected as a by-product. Examples include tea, coffee, rubber trees, oil and coconut palms and bamboo plantations, and tall grasses.

4   *Trees outside forests and woodlands.* These include bush trees, urban trees, roadside trees and on-farm trees. Trees outside forests have a major role as sources of fruits, firewood, etc., and their importance should not be underestimated.

5   *Agricultural crops.* These are crops grown specifically for food, fodder and fibre or energy production. Distinctions can be made between intensive, larger-scale farming, for which production figures may show up in the national statistics, and rural family farms, cultivated pasture and natural pasture.

6   *Crop residues*. Including crop and plant residues produced in the field. Examples include cereal straw, leaves and plant stems. Fuel switching can result in major changes in how people use biomass energy resources. For example, in China a rapid switch from agricultural residues to fossil fuels is causing serious environmental problems.

7   *Processed residues*. Includes residues resulting from the agro-industrial conversion or processing of crops (including tree crops), such as sawdust, sawmill off-cuts, bagasse, nutshells and grain husks. These are very important sources of biomass fuels and should be properly assessed.

8   *Animal wastes from both intensive and extensive animal husbandry*. When considering the supply of biomass it is important to ascertain the amount that is actually accessible for fuel, not the total amount produced. There are often large variations in assessments of animal wastes due to a lack of a common methodology, and wide variations in livestock type, location, housing and feeding conditions, etc. Animal waste may also have a better value as fertilizer. In addition, animal wastes are frequently used for producing biogas for environmental rather than energy purposes. Your survey needs to reflect the changing uses for animal wastes, and the reasons for these changes.

Classification into woody and non-woody biomass

Classification into woody and non-woody biomass is often for convenience only, as there is no clear-cut division between them. The way these biomass types are classified should not dictate which data is collected. For example, cassava and cotton stems are wood, but as they are strictly agricultural crops it is easier to treat them as non-woody plants. Bananas and plantains are often said to grow on banana trees, although they are also considered as agricultural crops. Coffee husks are treated as residues, whereas the coffee clippings and stems are classed as wood.

In some areas tall grasses are also used for energy (e.g. cooking and heating). More recently, various grasses (e.g. *Miscanthus*, elephant grass; see Chapter 4) have been investigated as possible sources of energy in modern commercial applications. In this case commercial measuring techniques would apply. Generally, the non-woody biomass includes the following:

- agricultural crops,
- crop residues,
- processing residues,
- animal wastes.

Woody biomass is perhaps one of the most difficult measurements to make, but it is usually the most important form of biomass energy to document, particularly when used in traditional applications. Past surveys have often concentrated on natural forests, plantations or woodlands, as many people have the idea that this is where fuelwood and timber come from. However, many demand surveys show that trees outside of forest or woodlands are a very important source of fuel, poles and even roughly hewn or hand-sawn timber. As these non-traditional sources of wood are so neglected, and because there is much more diversity of tree management in these areas, more effort should be put into measuring these trees than 'forest' trees. It may therefore be necessary to spend a great deal of your resources collecting data on woody biomass. Every effort should be made to determine growing stock and the annual growth increment of standing forests, in particular (see Chapter 3).

The methodology employed for estimating non-woody biomass depends on the type of material and the quality of statistical data available or deducible, and often the end use. For the purposes of policy making, an assessment of non-woody biomass can rely on statistics for agricultural production and any information available concerning agro-industrial residues. Such data will give a quick picture of the possible supply of such non-woody biomass resources as crops, crop residues and processing residues. If the required data is not available then an estimate of annual biomass production can be obtained from agricultural land use maps, in conjunction with reported figures on crop yield per unit area. Field studies and experimentation should be kept to the very minimum.

Where finer detail is required for the implementation of projects, exhaustive investigation is necessary to provide a clear, unambiguous report on each type of biomass resource, and to provide a detailed analysis of the availability, accessibility, collectability, convertibility, present use pattern and future trends.

For agricultural crops, dependable information on yields and stocks, quantified accessibility, calorific values, storage and/or conversion efficiencies must be accurately determined. A study of the socio-cultural behaviours of the inhabitants of the project area will help to determine use patterns and future trends.

In addition, residue indices for crop residues and processing residues must also be determined. It is also important to critically analyse the various uses of such residues (see Appendix 2.1).

The amount of woody biomass actually available for fuel also depends upon the alternative uses for biomass. Wood is rarely grown specifically for fuel, because fuelwood is cheaper (sometimes a great deal cheaper) than wood sold for other purposes. An alternative use normally has priority over use of wood as fuel.

However, when trees are converted to wood products there is considerable waste generated. In forests or on farms the buyer (or seller) only removes logs of specific dimensions. Branches and crooked stem wood may be left when the trees are felled, which can amount to between 15 and 40% or more of the above-ground volume. Sawn wood may only account for between one-third and one-half of original saw logs; that is, the waste is 50–67% of the log. All the waste materials are potentially burnable, especially if they are near to the site of demand. Even after wood is converted into, say, poles or sawn timber it can still be used for fuel after its useful life is over. All these facts have to be considered (again, see Appendix 2.1).

Proximity to demand centres or markets is also very important. The nearer it is to the consumer, the more likely it is that most if not all the biomass will be used. On the other hand, if the wood is remote from the demand there may be a surplus of biomass that is not readily available as fuelwood or charcoal. Thus, once a biomass supply map is drawn up it is important to match it with population densities to get a picture of how much of this biomass it is feasible to use.

Fuel switching is also very important. You need to assess the availability of other fuels. In a world that is urbanizing rapidly people are switching to more convenient fuels for economic and social and cultural factors if there are other, better alternatives, even if they are more expensive.

The importance of the agro-climatic zone

Biomass yield will vary with biomass type and species, agro-climatic region, rainfall and management techniques (e.g. intensive or extensive forestry or agriculture, irrigation and the degree of mechanization, etc.). A proper analysis of agro-climatic zones is therefore necessary for an accurate assessment to be carried out. If this information is not available, some form of remote sensing is the most viable option (see Chapter 8). The particular technique chosen will depend on local circumstances.

Variation with agro-climatic zone and with time

The productivity of biomass will probably vary in different seasons and over years. Time-series data should be collected whenever possible. Particular care is needed if the biomass is not homogeneous. When estimates of vegetation cover are made in the absence of detailed land use data it is important to recognise the effect of climate, altitude and geology on plant growth. Water is a key factor in productivity, so it is important to get figures for rainfall over time.

Assessment of accessibility

The potential supply and the actual availability of biomass are very rarely the same. The proportion of the total biomass that is accessible, and therefore available for fuel, is one of the hardest and most essential tasks in any study of biomass supply. The variability and complexity of the factors involved with access can make quantitative analysis difficult. Access to potential biomass resources is limited by three main physical and social constraints:

1   location constraints,
2   tenure constraints (social, political and cultural),
3   constraints derived from the system of land resource management.

Locational constraints

Locational constraints reflect the physical difficulties of harvesting, collecting and transporting biomass from the point of production to the place where it will be burned. The gathering and transport of biomass is influenced by the terrain and the distance over which the biomass is transported, and also by the availability of biomass in a determined area. Rivers, steep slopes and areas of marshland and so on all act as barriers to access. Locational constraints greatly influence the cost of biomass energy in both labour and financial terms.

It is possible to assess localized constraints cartographically, by measuring the distance between resource and consumers and noting features of the terrain. If required, more detailed information may be obtained from detailed cartographic data (from existing maps and/or a remote sensing analysis). Fieldwork will provide further information on the time necessary to collect fuel, distances travelled and the points at which time and distance begin to limit accessibility. Locational constraints are important, and need careful analysis.

Tenure and land management constraints

Tenure and land management constraints stem from land ownership/land rights. Land tenure patterns are highly specific to individual countries, and political and cultural aspects are very important. It is possible to identify three broad categories of land ownership:

1   small farms owned or rented by people in the local community, where resources are subject to private property rights,
2   communal land, owned by local people or groups or local communities,
3   large areas of land owned and controlled by individual landowners and institutions, either commercial farms and plantations or state land such as forest and game reserves.

Issues of tenure do not always affect small farms and communal land unless there are major political and social changes occurring. Thus an assessment of tenure constraints should concentrate on areas where land is held by the state and the commercial farming and plantation sector. Where plantations, state reserves and the like are frequently closed off to the local community, information on accessibility is easily obtained from the institution controlling the area. Reference to these institutions will establish the policies regarding access by the local community for fuelwood collection. Where the land is owned or used by many small farmers, the task will be more complex since many people may have to be contacted to gain access.

The question as to whether private property rights on small farms constrain accessibility is often difficult to resolve. It may be hard to determine whether a resource is limited because the supply is inaccessible, or because certain sections of the community are deprived of access. Here your social and political skills will be needed! If a detailed analysis is required, some level of fieldwork is needed to establish additional factors, such as the extent of illicit collection.

Stock and yield

The assessment of all biomass resources (woody, non-woody and animal) requires the measurement estimation of both stock and yield. If biomass is viewed as a renewable resource, it is the annual production, or increment, that is the key factor. Stocks become depleted when the biomass harvested is greater than the increment.

Stock is defined as the total weight of biomass as dry matter. *Yield* is defined as the increase in biomass over a given time and for a specific area. Yield must include all biomass removed from the area. Yields are expressed in two forms, as follows.

1   *Current annual increment* (CAI): the total biomass produced over a period of 1 year. For annual plants, it is the total yield over the year. For perennial plants, such as trees and other woody biomass, the CAI will vary according to the season and the growing conditions. For perennials, a mean figure should be calculated from measurements made every year and in the same season.
2   *Mean annual increment* (MAI): the total biomass produced for a certain area divided by the number of years taken to produce it. MAI is an *average* measure of yield (see Glossary, Appendix 10.1).

Animal stock is measured as the number of animals by species.

Having determined the quantity of biomass potentially available, its energy value will depend on moisture and ash content.

Moisture content

When biomass is burned, part of the energy released is used to turn the water it contains into steam. It follows that the drier the biomass, the more energy there is available for heating. It is therefore the moisture content that primarily determines the energy value of the biomass. Thus, while wood has a higher energy value than the other two forms of biomass at a given moisture content, it is possible for crop residues and dung to have a higher value than wood if they have lower moisture contents.

The energy value of a unit weight of biomass is inversely proportional to the amount of water it contains. To get the true weight of biomass it is therefore necessary to calculate the moisture content (mc). This can be measured in two ways: on a wet basis (wb) or dry basis (db). These measurements are calculated as follows.

- Dry basis (db)

$$\text{Moisture content} = \frac{\text{Wet weight} - \text{dry weight}}{\text{Dry weight}}$$

- Wet basis (wb)

$$\text{Moisture content} = \frac{\text{Wet weight} - \text{dry weight}}{\text{Wet weight}}$$

Air–dry wood (15% mc db) has an energy value of about 16.0 MJ/kg, whereas green wood (100% mc db) has a value of 8.2 MJ/kg. The energy value of oven dry woody biomass can be taken as 18.7 MJ/kg, ± 5%. Resinous wood has a slightly higher value, and temperate hard wood a slightly lower value.

Ash content

The higher the ash content, the lower the energy value. On an 'ash-free' basis – that is, when the non-combustible material is discounted – all non-woody plant biomass has more or less the same energy value.

Ash contents vary from one crop/residue to another. For example, rice husks have 15% ash content, maize cobs 1%, so they have different energy values. Bone-dry rice husks with a moisture content of 15% are 85% fibres, but as 15% of this is not combustible, the husks consist of 70% combustible material. Maize husks on the other hand, at 15% mc db, are 84% combustible material, 15% more than the rice husks.

Ash contents should be compared between biomass samples with the same moisture content.

Energy value of biomass

Further details are given in Appendix 2.2.

The energy available from biomass is expressed in two main forms:

1 *gross heating value* (GHV), also expressed as higher heating value (HHV); and
2 *net heating value* (NHV), also called lower heating value (LHV).

Although for petroleum, for example, the difference between the two is rarely more than about 10%, for biomass fuels with widely varying moisture contents the difference can be very large and therefore it is very important to understand these parameters.

GHV refers to the total energy that would be released through combustion divided by the weight of the fuel. It is widely used in many countries. The NHV refers to the energy that is actually available from combustion after allowing for energy losses from free or combined water evaporation. It is used in all the major international energy statistics. The NHV is always less than the GHV, mainly because it does not include two forms of heat energy released during combustion: (1) the energy to vaporize water contained in the fuel and (2) the energy to form water from hydrogen contained in hydrocarbon molecules, and to vaporize it.

Calculating energy values

For zero-moisture wood the energy value is 20.2 MJ/kg, for crop residues it is 18.8 MJ/kg and for dung it is 22.6 MJ/kg. The difference between HHVs and LHVs is approximately 1.3 MJ/kg at 0% mc. This figure is therefore deducted from the HHVs of 18.9 MJ/kg for wood, 17.6 MJ/kg for crop residues and 21.3 MJ/kg for dung to obtain the LHV and to calculate the moisture and ash contents. Wood with a moisture content of 80% db contains 44% water and 56% fibre. If all the fibre is burnable, the energy content is $0.56 \times 18.9 = 10.6$ MJ. However 1% is non-combustible, so the energy content is $0.56 \times 18.9 \times 0.99 = 10.5$ MJ/kg. Some of this energy is required to drive off the water. To expel 0.44 kg of water will take $2.4 \times 0.44 = 1.1$ MJ (heat required to drive off 1 kg of water). Thus the net energy available for heating is $10.5 - 1.1$ MJ/kg. This formula can be used if the HHV or the LHV is known at specific moisture and ash contents (see Appendix 2.2).

Furthermore, the difference between NHV and GHV depends largely on the water (and hydrogen) content of the fuel. Petroleum fuels and natural gas contain little water (3–6% or less) but biomass fuels may contain as much as 50–60% water at the point of combustion. Heating values of biomass fuels are often given as the energy content per unit weight or volume at various stages: green, air-dried and oven-dried material (see Glossary, Appendix 10.1).

Many surveys go into elaborate detail about energy values of fuel and record them to several decimal places. As data is usually only accurate to within at best 20%, such detail is generally unrealistic. The net amount of energy available from biomass as heat depends upon two factors:

1 the amount of water it contains,
2 the quantity of non-combustible material in the biomass.

The net amount of energy available for heat is the total mass of the material in question, taking into account the amount of water it contains and the quantity of non-combustible material which will be left as ash after burning (the ash content). The substances that form the ashes generally have no energy value. For woody biomass the ash content is more or less constant at around 1% for all species. For non-woody biomass, ash content can be more important.

It is therefore the moisture content rather than the species of wood that determines energy availability.

Energy values of crop residues

For further details see Appendix 2.1.

On an ash-free basis, the energy value of crop residues is slightly less than that of wood, principally because they have a lower carbon content (about 45%) and higher oxygen content.

The average energy value of ash free, bone-dry annual plant residues is about 17.6 MJ/kg. At 15% mc db, the energy value of the ash free residue is about 15.0 MJ/kg. With a 2% ash content, the energy value will be about 14.7 MJ/kg, and with a 10% ash content, about 13.5 MJ/kg.

Dung

The ash-free energy value of animal dung is higher than that of wood. Bone-dry dung has a LHV of about 21.2 MJ/kg. At 15% mc db this is reduced to 18.1 MJ/kg. However, with an average ash content ranging from 23 to 27%, the actual energy value at 15% mc db is about 13.6 MJ/kg.

Charcoal

The energy value of charcoal depends not only on its moisture and ash contents, as for other forms of biomass, but also on its degree of carbonization. The average moisture content of charcoal is about 5% (db) and its ash content depends on the parent material. Wood charcoal may have up to 4% ash content and coffee husk charcoal 20–30%. Charcoal is obtained by the carbonization (pyrolysis) of wood by heat in the absence of air at a temperature above 300°C, when volatile components of wood are eliminated. During the process there is an accumulation of carbon in the charcoal from about 50% to about 75%, due partly to the reduction of hydrogen and oxygen in the wood. The assessment of various aspects related to charcoal is dealt with in detail in Chapter 3.

Weight versus volume

The forest industry measures wood by volume, but biomass fuels should be measured by weight. This is because the heating value, or amount of heat that can be provided, must be referred to on a weight basis. Weight estimates can be obtained directly from measurements of tree dimensions, or indirectly via wood volume measurements. Direct determination of weight is preferable for biomass energy supply assessment.

Biomass that is traded – e.g. timber and agricultural commodities – is measured in standard units suited to a particular commodity. For example, foresters measure timber by volume because they are concerned with the bulky and more-or-less uniform stems and trunks. However, biomass fuels are often irregularly shaped (as twigs, small branches, leaves, stalks, etc.), making volume an awkward way of measuring biomass. In addition, the weight is required when determining the heat value of biomass. Biomass for energy should always be measured by weight.

Estimating biomass flows

A long-term goal in data collection is to produce a complete biomass flow diagram for the country, region or locality. A flow diagram traces biomass from production to end use. It should encompass all forms of biomass production (agriculture, forestry, grasslands, etc.), allow for losses during conversion and provide details of all its uses (food, feed, timber, fuelwood, animals). But this can be both time-consuming and expensive.

To construct a flow diagram, data has to be collected systematically, starting with aggregated data and working towards a fine level of detailed information. As flow diagrams are drawn to scale, units must be consistent throughout. Accurate flow diagrams therefore require a painstaking analysis of biomass supply and consumption. However, once established, such a diagram is a useful method of presenting data, can give an excellent national, regional and local overview, and provide an easy means of monitoring changes in the production and use of biomass. Biomass energy flow diagrams are dealt in detail in the Case study 9.3.

Thinking specifically of setting up a modern biomass energy project?

If you wish to set up biomass energy project (biomass-based power plant), here are some specific ideas. You need to investigate in detail all the pros and cons, not least the sustainability implications (see Chapter 6). You need to pose a few questions – taking into account not only the nature of biomass for energy, but also methodologies for assessing production, supply and demand – and then answer them, such as: competing land uses, how to develop energy plantations, sustainability and certification issues, soil nutrients and water impacts, impact on biodiversity, transport costs and environmental impacts. More details follow.

To set up a biomass project, then, can be complicated. For example, you need to ensure:

- how much biomass is available during the duration of the project, and sustainably;
- how to deliver the feedstock to the plant: given its bulky nature and energy density, this needs to be carefully assessed;
- quality and moisture contents: these are very important, so do not overlook them;
- if the feedstock needs to be imported, or transported long distances, then consider not only the costs but also sustainability, certification, etc. (Chapter 6). This can be an important component of the final cost;
- size of the plant and nature of the conversion technology varies in function of the nature of biomass;
- availability of the resources (what type of biomass);
- potential long-term impacts;
- planning issues (very important), often requiring many legal contracts which can be time-consuming and costly;

- competing uses of biomass: the price of the feedstock will depend on its overall availability and diversity of end uses;
- environmental impacts, which are extremely important (see below);
- social impacts: often overlooked in past, but no longer the case (see below);
- the level of investment needed (see below).

Ensuring you have sufficient biomass to run your plant can be a huge task and you need to pose many questions, as follows.

- Who is going to produce the biomass envisaged for biofuel production? Although this market is maturing rapidly (see www.bioenergytrade.org) it is still in its formative stage and many gaps remain in our understanding of how this market operates.
- Can profitable biomass production systems be developed that 'fit' with processing and markets for the kinds of final products? What form will these take, and what investment will be required?
- How can the environmental benefits from biomass production be maximized? For example, how to get the best net energy or carbon balance per hectare, or per person day?

Table 2.1 summarizes the scale of operation of a typical bioenergy plant.

A major difficulty in the development of biomass power plants, often overlooked by the academic community, is how to convince potential investors. To gain their confidence a detailed financial analysis needs to be prepared.

Table 2.1 Typical bioenergy plant scale-operations (IEA, 2007)

Type of plant	Heat or power capacity ranges, and annual hours of operation	Biomass fuel required (oven-dry tonnes/year)	Vehicle movements for biomass delivery to the plant	Land area required to produce the biomass (% of total within a given radius)
Small heat	100–250 kW$_{th}$ 2,000 h	40–60	3–5/year	1–3% within 1 km radius
Large heat	250 kW$_{th}$–1 MW$_{th}$ 3,000 h	100–1,200	10–140/year	5–10% within 2 km radius
Small CHP	500 kW$_e$–2 MW$_e$ 4,000 h	1,000–5,000	150–500/year	1–3% within 5 km radius
Medium CHP	5–10 MW$_e$ 5,000 h	30,000–60,000	5–10/day	5–10% within 10 km radius
Large power plant	20–30 MW$_e$ 7,000 h	90,000–150,000	25–50/day and night	2–5% within 50 km radius

Source: © OECD/IEA (2007), *Good Practice Guidelines: Bioenergy Project Development & Biomass Supply*, IEA Publishing. Licence: www.iea.org/t&c/termsandconditions, modified by Routledge Books | Taylor and Francis Group.

- You need to prepare a detailed assessment of biomass resource potential.
- Ask, are there sufficient resources available on a long-term and sustainable basis?
- Has a detailed cost-benefit analysis been done?
- What type of biomass resources are available locally?
- What are other uses of biomass resources other than for energy? What are the possible long-term competitive uses for biomass other than for energy?
- Are there other competitive renewable energy alternatives, e.g. wind?
- What is (or is likely to be) the long-term government policy towards bioenergy?
- What are the subsidies, including fossil fuels, if any?
- What are people's social attitudes to biomass plants?
- Consider carefully the benefits of stand-alone plants versus combined use; e.g. use in co-firing.
- What is the role of energy decentralization and what place does biomass energy have in this context?

A major stumbling block for large-scale biomass power plants is how to ensure the continuous and sustainable supply of the feedstock and, in the case of plantations, there are some hard questions that you need to pose:

- How high are the yields, and how sustainable are they?
- What is the availability of large amounts of biomass?
- What are the costs?
- What is the quality of the biomass?
- Is there security of supply?
- What is the energy efficiency of your conversion technology? Could this be improved?

From the outset, you also need to add non-physical factors to the list, including the following.

- Direct public participation is a key issue. It is important to involve the local community in any project from the outset.
- Are there any transport, distribution or communication issues?
- What is the current company policy on your enterprise (particularly long-term prospects)?
- What is the regulatory framework? Do you need to improve it?
- Marketing: what is the current size of your market and what is the potential for growth?
- Consider consumer awareness/acceptance.
- Consider potential customers' understanding of bioenergy.

The implications of social participation are major issues you have to consider in detail. For example:

- Is there a poor perception of biomass energy, e.g. use of outdated technology, pollution, low efficiency?
- Will there be any impacts on biodiversity?
- Will there be any impacts on forests, e.g. clearing?
- Will there be any impacts on local transport?
- Will there be impacts on water use?
- Are there any planning issues?

Also, do not underestimate the power of local authorities and social groups. You may have to deal with a large number of issues:

- policy planners,
- local government agencies,
- electricity regulators/utilities,
- local pressure groups,
- health and safety,
- land owners/farmers,
- wildlife,
- conservation groups.

Often biomass energy projects get a bad name, primarily because they have overlooked social impacts, which play a growing role:

- child labour,
- freedom of association,
- discrimination,
- forced labour,
- wages,
- working hours,
- contracts and subcontractors,
- land rights.

And finally, the implications for certification and sustainability are currently at the core of biomass energy plants. This involves additional costs which are not as yet fully quantified (see also Chapter 6).

Future trends

You must be aware of changing trends in measuring techniques, fuel switching, social, economic and policy changes, potential energy alternatives and ways and means for enhancing existing resources in an environmentally sustainable manner. Environmental implications have so often been ignored in the past, and must be fully taken on board.

It is also essential to be aware of the potential for increasing yields from existing and new species or clones. This knowledge should be based on national and international research and experience. You must also appreciate the conflicting pressure of land use trends and socio-economic changes that will affect the possibility of maintaining (or increasing) biomass supplies. Such information needs to be continuously made available at the planning and policy levels, otherwise the 'on-the-ground' decisions may be very difficult to implement.

The aim is to achieve optimal and sustained production of biomass in a manner that fulfils environmental, socio-economic and sustainability criteria, to allow the policy maker to make sound policy decisions.

References and further reading

BIALY, J. (1986) *A new approach to domestic fuelwood conservation: guidelines for research*. FAO, Rome.

CARPENTIERI, A.E., LARSON, E.D. AND WOODS, J. (1993) Future biomass based electricity supply in Brazil. *Biomass & Bioenergy* 4(3), 149–176.

FAO (ND) *Statistics at FAO*. www.fao.org/waicent/portal/statistics_en.asp.

FISCHER, G., SHAH, M., VELTHUIZEN, H. AND NACHTERGAELE, F. (2006) *Agro-ecological zones assessment*. Interim Report RP-06-003. International Institute for Applied Systems Analysis (IIASA), Luxembourg.

HALL, D.O. AND OVEREND, R.O. (eds) (1987) *Biomass: regenerable energy*. John Wiley & Sons, Chichester.

HALL, D.O., ROSILLO-CALLE, F. AND WOODS, J. (1994) Biomass utilization in households and industry: energy use and development. *Chemosphere* 29(5), 1099–1119.

HEMSTOCK, S.L. (2005) *Biomass energy potential in Tuvalu*. (Alofa Tuvalu) Government of Tuvalu Report.

HEMSTOCK, S.L. AND HALL, D.O. (1994) A methodology for drafting biomass energy flow charts. *Energy for Sustainable Development* 1, 38–42.

HEMSTOCK, S.L. AND HALL, D.O. (1995) Biomass energy flows in Zimbabwe. *Biomass and Bioenergy* 8, 151–173.

HEMSTOCK, S., ROSILLO-CALLE, F. AND BARTH, N.M. (1996) BEFAT - Biomass Energy Flow Analysis Tool: a multi-dimensional model for analysing the benefits of biomass energy. In *Biomass for energy and the environment*, Proc. 9th European Energy Conference, Chartier et al. (eds). Pergamon Press, pp. 1949–1954.

IEA (2007) *Good practice guidelines: bioenergy project development and biomass supply*. IEA, Paris. www.iea.org/publications/freepublications/publication/biomass.pdf.

KARTHA, S., LEACH, G. AND RJAN, S.C. (2005) *Advancing bioenergy for sustainable development: guidelines for policymakers and investors*. Energy Sector Management Assistance Programme (ESMAP) Report 300/05. The World Bank, Washington DC.

LEACH, G. AND GOWEN, M. (1987) *Household energy handbook: an interim guide and reference manual*. World Bank technical paper no. 67. World Bank, Washington DC.

OGDEN, J., WILLIAMS, R.H. AND FULMER, M.E. (1991) Cogeneration applications of biomass gasifier/gas turbine technologies in cane sugar and alcohol industries. In *Energy and the environment in the 21st century*, J.W. Tester, D.O. Wood and N.A. Ferrari (eds). MIT Press, Cambridge, MA, pp. 311–346.

ROSILLO-CALLE, F. (2001) Biomass. In *Biomass energy (other than wood) commentary 2001.* World Energy Council, London, pp. 254–264.

ROSILLO-CALLE, F., FURTADO, P., REZENDE, M.E.A. AND HALL, D.O. (1996) *The charcoal dilemma: finding sustainable solutions for Brazilian industry.* Intermediate Technology Publications, London.

ROSILLO-CALLE, F., DE GROOT, P., HEMSTOCK, S. AND WOODS, J. (eds) (2006) *The biomass assessment handbook: energy for a sustainable development.* Earthscan, London.

SMITH, C. (1991) Rural energy planning: development of a decision support system and application in Ghana, PhD thesis, Imperial College of Science, Technology and Medicine, University of London.

WORLD RESOURCES INSTITUTE (1990) *World resources 1990–1991: a guide to the global environment.* Oxford University Press, Oxford, UK

Appendix 2.1   Residue calculations

J. Woods

Forestry

1   Data from the 1989 *FAO Forest Products Yearbook* calculated solely from 1988 'roundwood production' figures.

2   It is assumed that 'roundwood' (which is synonymous with previously defined 'removals' in prior Yearbooks) is equivalent to 60% of the total volume of wood actually cut (i.e. total cut equals 1.67 times roundwood production).

3   The 60% figure is based on the amount of commercial stem wood which is available from the total tree above-ground biomass (see Hall and Overend, 1987); thus only the stem and large branches are removed from the cutting site.

4   On a global basis, of the wood removed from site ('roundwood'), approximately half is used for 'industrial roundwood' and the remaining half for 'fuelwood and charcoal' (data from the *Yearbook*). Historically, at least 50% of the 'industrial roundwood' was 'lost' (predominantly as sawdust) during processing, most of which could be considered as potentially harvestable residues. The amount of wood which remains as residues varies widely and depends mainly on process efficiency and economics. More recently, and particularly in Organisation for Economic Co-operation and Development (OECD) countries, residues generated at the mill are at least partially used for other purposes, predominantly the manufacture of particleboard and MDF (medium density fibreboard). Work is being carried out to quantify the impacts of the improvements in utilization of the wood residue fraction in mills.

5   'Potentially harvestable residues' includes all on-site forestry residues (r1, i.e. 40% of total cut wood) plus all residues arising from 'industrial roundwood' processing at the timber mills (r2, i.e. 50% of 'industrial roundwood,' calculated for each country). See Figure 2.2. Practically, we assume that only 25% of the potentially harvestable residues are 'recoverable'.

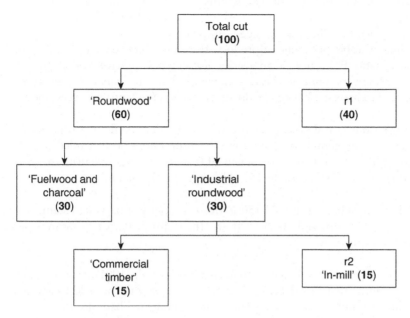

Figure 2.2 Diagrammatic breakdown of forestry production to show origin of residues: a
hypothetical example (based on global averages).

Notes: total cut = 1.67 * 'roundwood'. 'Industrial roundwood' = 'roundwood' –
'fuelwood and charcoal'. r1 = portion of total cut left on site, equivalent to 40%
of the total cut; r2 = portion (50%) of 'industrial roundwood' lost through processing. 'Potentially harvestable residues' = r1 + r2 = 55% (globally). Please note
that substantial fractions of 'r2' are now used by the forest industry for particle
board, fibreboard and other types of construction materials. Forestry residues
were analysed individually to deduce r1 and r2. Individual regional totals are
aggregates of the individual countries.

Decisions as to the amount of residues that should be left on site for soil conditioning, nutrient recycling, etc., will vary greatly for any particular site.
However, in our final calculation of biomass energy supplies in by country or
region, only 25% of the potentially harvestable residues are used to calculate
the 'recoverable residues'. This should allow sufficient leeway for sustainable
residue collection, but this can only ever be an informed decision made for
each individual site, arrived at by taking fully into consideration all site–specific
factors and preferably including ongoing monitoring.

Crop residues

1   Data is from *FAO Production Yearbook*, 1989, for cereals, vegetables and
    melons, and sugar cane. Roots and tubers and sugar beet data is from the
    FAO's Agrostat database.

2   Since comprehensive global data is only available for crop production the 'potentially harvestable residue' resource is estimated using residue production coefficients. These allow rough estimates to be made of the amounts of residues available per tonne of product; thus it is effectively a by-product to product ratio. For cereals, as an average for all types we have used a figure of 1.3, i.e. for every tonne of wheat, corn, barley grain, etc., harvested there is the potential to harvest 1300 kg (air-dried) of residues ('potentially harvestable').

Sugar cane residues are also calculated from FAO production figures that provide the amount of cane stems harvested. Using the assumptions of Alexander (quoted in Carpentieri, 1993) a residue coefficient of 0.55 was used to calculate the residue production figures (see Table 2.2). This figure is derived from:

• production of bagasse at 0.3 t (50% moisture) per t of harvested stem,
• use of 'green tops and attached trash' (barbojo) at 0.25 t (50% mc) per harvested t of stem.

The barbojo figure is lower than is used in Ogden et al. (1991) (0.66 t/t cane stem); we assume that the 'detached leaves' will remain in the field to act as a soil conditioner and are not a component of the collected barbojo.

3   We use a coefficient of 1.0 for vegetables and melons, which although seemingly high, may be justified by a study which showed, in an extreme example, that only about 6–23% of the harvested lettuce plant is finally eaten depending on the season There are other problems in dealing with factors such as changing water content, realistic levels of collectability for energy use, and amounts used as animal feed. However, until better data is available we will continue to use this deliberately simplistic coefficient. It should also be noted that vegetables + melons, roots + tubers, and sugar beet together only provide 3% of the total 'potentially harvestable residues' while cereals provide 32% and sugar cane 6% of the total.

4   As with forestry residues, the amount of residues which can be removed from the field relies on a wide variety of factors that are site-dependent.

Table 2.2 Crop residue production ratios

Crop	Production coefficients		Energy content	
	t/t	*Moisture*	*GJ/t (HHV)*	*Moisture*
Cereals	1.3	Air dry	12	Air dry
Vegetables and melons	1.0	Air dry	6	Air dry
Roots and tubers	0.4	Air dry	6	Air dry
Sugar beet	0.3	Air dry	6	Air dry
Sugar cane	0.55	50%	16	Air dry

However, we use only one-quarter of the calculated 'potentially harvest-able residues,' and call these 'recoverable residues'. Thus we allow for the worst cases where most of the residues will be required for protection against erosion, nutrient recycling and water retention.

5   The total energy content for 'potentially harvestable' crop residues (15.59 EJ developed, 21.51 EJ developing and a total of 37.10 EJ for the world) does not include estimates for residues from pulses, fruits and berries, oil crops, tree nuts, coffee, cocoa and tea, tobacco or fibre crops. Although regional estimates of their recoverable residue potential would be too small to appear in these tables they may still be a locally significant energy resource, but will not often figure significantly on a national scale.

Dung

1   Dung production is calculated using data from FAO and UN Population Division (World Resources Institute, 1990). This data only shows the numbers of animals, disaggregated into different commercial species, and it is therefore necessary to use the dung production coefficients shown below to estimate the total energy potential from animal dung production.

2   Dung production coefficients are shown in Table 2.3.

3   Due to the dispersed nature of dung, it is estimated that only 50% of the dung actually produced is 'potentially harvestable'; we consider that only 25% of the 'potentially harvestable' dung is 'recoverable,' thus only one-eighth of the total dung actually produced by the animals is calculated as 'recoverable dung'. It should be noted that in a number of developing countries dung already plays a significant role in the domestic energy sector, e.g. India and China.

Table 2.3 Dung production coefficients

	Dung production rate	Energy content
	[kg (oven dry) per day per animal]	(GJ/t, oven dry)
Cattle	3.0	15.0
Sheep and goats	0.5	17.8
Pigs	0.6	17.0
Equines	1.5	14.9
Buffaloes and camels	4.0	14.9
Chickens	0.1	13.5

Sources: derived from Taylor, T.B., Taylor, R.P. and Wiess, S. (ND) Worldwide data related to potentials for widescale use of renewable energy. Report no. pu/cees 132. Center for Energy and Environmental Studies, Princeton University, NJ and Senelwa, K.A. and Hall, D.O. (1993) A biomass energy flow chart for Kenya. *Biomass and Bioenergy* 4, 35–48.

Note: the coefficients are predominantly based on developing country dung production rates and may significantly underestimate production rates in OECD countries.

Appendix 2.2   Volume, density and moisture content

Jeremy Woods and Peter de Groot

Volume

Woody biomass, particularly fuelwood, production and consumption are normally measured by volume. But frequently in informal markets and household surveys the only record of fuelwood quantities produced, sold or consumed is a volume measure based on the outer dimensions of a loose stack or load containing air spaces between the wood pieces, such as the stere, cord, truckload, headload or bundle.

Two approaches can be employed to use such measures for energy analysis: (1) converting stacked volume to a weight, e.g. by weighting a number of examples with a spring balance (small load) or a weighbridge (truckload) or (2) converting stacked volume to solid volume, e.g. by immersing loads (small ones) in water and then measuring the volume of water displaced.

However, if it is not possible to convert volumes to weights, then the volumes of fuels have to be converted to a volumetric measure. To achieve this, a series of three conversions is normally required, as described below.

Conversion of oven-dry volume to oven-dry weight

$$\text{Oven} - \text{dry weight} \left(\text{ODW} \right) = \text{volume} \left(\text{m}^3 \right) \times \text{basic density} \left(\text{kg}/\text{m}^3 \right)$$

and since

$$\text{Basic density (kg/m}^3) = \text{specific gravity of dry matter}$$
$$(\text{g}/\text{cm}^3 \text{or t}/\text{m}^3) \times 1000 \ (\text{g}/\text{kg or kg}/\text{t})$$

then

$$\text{Oven} - \text{dry weight} \left(\text{ODW} \right) = \text{Volume} \times \text{Specific gravity} \times 1000$$

Conversion of oven-dry weight to actual weight for specific moisture content

$$\text{Actual wet weight} = \text{ODW}/(1 - \text{W}/100)$$

where W is the percentage moisture content, wet basis (mc wb).

Conversion of actual wet weight at specific moisture
content to net heating value given the oven-dry value

Use actual weight and the formulae given in the main part of the chapter for heating value per unit weight. These formulae can be combined to give a single formula for converting from volume (V), basic density (BD), oven-dry gross heat value (Z) and a percentage moisture content wet basis (W) to the net heating value (NHV), as recommended and used in this book.

$$\text{NHV} = V \times BD \times [Z - 1.3 - (W/100) \times (Z + 1.1)]$$

(of given volume) $1 - W/100$
    where volume is in cubic metres, weight is in kilograms and energy is in mega-Joules.
    The starting point of the chain (in this sample) is one solid cubic metre of green wood at the point of harvest weighing 1265.8 kg/m^3 and the oven-dry energy value is 20 MJ/kg.

Moisture

See main chapter text for more details about moisture. It is important to know the moisture content of wood since this can have important effects, particularly if weight is the measure for assessing the solid volume, and the energy content. For example, if wood is cut green and then collected, the moisture content could be 100% or more (dry basis). If the wood is first allowed to dry out (air dry) then the moisture content may be around 12–15%, depending on the relative humidity of the atmosphere. While the energy given off from a piece of wood is about the same on a weight basis irrespective of species, the moisture content of wood determine the heat value of wood, e.g. 1 kg of wood at 15% moisture content when burned will give off about 16.0 MJ, and with 40% moisture content the heat value is about 12.7 MJ/kg.
    In this section all moisture contents have been given on a dry basis but in some countries fuelwood moisture content is given on a wet basis. The formula for changing from dry to wet basis is as follows:

$$D = W/[(1-W)/100]$$

where D is moisture content as a percentage of the dry weight basis and W is moisture content as a percentage of the wet weight basis.
    If the moisture content is 100% on the dry basis then the wet basis moisture content is 50%. Similarly, a 15% dry–basis moisture content = 13% (wet basis).
    The reverse formula for changing from wet basis moisture content to dry basis moisture content is as follows:

$$W = D/[(1+D)/100]$$

Density

If using weight as a measure to determine solid volume, density becomes an important factor since weight depends on density and densities within and between species vary. For example wood from young trees is less dense than that from old tress of the same species, and sapwood is less dense than heartwood.

Further reading

KARTHA, S., LEACH, G. AND RJAN, S.C. (2005) *Advancing bioenergy for sustainable development: guidelines for policymakers and investors.* Energy Sector Management Assistance Programme (ESMAP) Report 300/05. The World Bank, Washington DC.

LEACH, G. AND GOWEN, M. (1987) *Household energy handbook: an interim guide and reference manual.* World Bank technical paper no. 67. World Bank, Washington DC., pp. 16–20.

OPENSHAW, K. (1983) Measuring fuelwood and charcoal. In *Wood Fuel Surveys.* FAO, Rome, pp. 173–176.

3 Assessment methods for woody biomass supply

Frank Rosillo-Calle, Peter de Groot,
Sarah L. Hemstock and Jeremy Woods

Introduction

Chapter 2 described the general methodologies for biomass assessment. This chapter looks at the most important methods for accurately measuring the supply of woody biomass for energy, and in particular techniques for:

- forest mensuration,
- determining the weight and volume of trees,
- measuring the growing stock and yield of trees,
- measuring the height and bark of trees.

And the energy available from:

- dedicated energy plantations,
- agro–industrial plantations,
- processed woody biomass (woody residues, charcoal).

Requirements prior to an assessment of biomass supply

Assess the natural resources

An analysis of biomass energy resources requires accurate assessment of a country's natural resources. These should include:

- land type,
- vegetation type,
- soil composition,
- water availability,
- weather patterns.

Box 3.1 illustrates a possible decision-making process when deciding the necessity for and detail of a field survey. Key issues to keep in mind are to be clear where you wish to focus your assessment, who is your audience, as stated in Chapter 2, and deciding whether the survey is necessary. Only consider mounting a field survey if there is no alternative, or if the areas involved are small.

Box 3.1 A decision tree for a formal woodfuel survey

1 Is problem defined, and scale and required precision established?

If not, stop!

2 Check existing information.

If adequate, stop!

3 Check redefine and clarify problem.

Repeat step 1.

4 After completion of steps 1–3 does existing knowledge need updating?

If not, stop!

5 Is Rapid Rural Appraisal (RRA) more suitable?

If yes, conduct RRA, and complementary focused small surveys.

6 Is there sufficient information to stratify population?

If not, go back to step 2. If still not consider stopping! Otherwise prepare steps 1–5.

7 Estimate sample size and budget.

If inadequate, repeat steps 1–5.

8 Are trained enumerators and data processors available?

If not: could they be trained? If not, repeat steps 1–5.
    If yes, commence training/recruitment (concurrent with steps 9 and 10).

9 Design draft questionnaire.
10 Is pilot testing OK?

If not, repeat steps 1–5.

11 If funds are available and staff are trained, and timing and seasonality
    are OK, commence fieldwork and related supervision.

If unsatisfactory, stop or return to step 1.

12 Data processing: checking, editing, coding entry and validation.

If unsatisfactory, correct and go to 13. If not corrected, stop.

13 Conduct statistical analysis and prepare tables of results.
14 Compare with initial hypothesis, previous results, local expert opinion and studies from other countries.
15 Return to step 1. Is further information still required?

If not, submit report.

16 Conduct follow-up studies, RRA, debriefing of field staff and spot surveys as required.

Submit report.

Measuring biomass variation in an agro–climatic zone over time

Biomass yield varies with the:

- type of biomass and the species,
- agro–climatic region,
- management techniques employed in biomass production (e.g. intensive or extensive forestry or agriculture, irrigation, the degree of mechanization, etc.).

These factors are important when making general estimates in the absence of detailed land use data. Furthermore, the productivity of biomass will almost certainly vary across seasons, and over years. It is important to collect time-series data wherever possible. Particular care is needed if the biomass is not homogeneous. Also, keep in mind that a lot of data is often available so do a literature review before starting your assessment. Pay particular attention to unrecorded removals from forests, trees on roadsides and community and farm lands, shrubs and scrub in the forest and branch wood because they are potentially large sources of energy and are often ignored or underreported.

Remote sensing

Remote sensing (dealt with in detail in Chapter 8) is an analysis of detailed land use patterns based on aerial photography or high-resolution satellite imagery.

The assessment of land use

This is assessed in greater detail in Chapter 7. Suffice to say that the method by which land use is assessed will depend upon the level of information required but it is advisable to begin collecting data at the most general, aggregated level suited to policy-level analysis. Further enhancement and disaggregation will almost certainly then be needed. A proper analysis of agro-climatic zones will be necessary; if this information is not available some form of remote sensing is the most viable option. The particular technique chosen will depend on local circumstances. It is therefore valuable to examine the change in land use over time. This can be done in two complementary ways:

- the analysis of historical remotely sensed data if this is available,
- through detailed discussion in the field with the local population.

The accessibility of biomass

Abundance of biomass *per se* can mean very little if it is difficult or costly to access, and hence its availability and the potential supply are rarely the same. It is therefore very important to take into account the actual accessibility of biomass, although this is often very difficult to measure. There are two main physical and social constraints that restrict access to biomass: locational constraints and tenurial constraints.

Key steps

An assessment of the supply of biomass involves the following key points.

- *Decide on the biomass assessment method most suited to your needs.* The method of estimating woody biomass, whether directly or indirectly, is determined by a number of factors, including the area in which it is growing, its variability and its physical size. You will probably have to put more effort into measuring scattered non-forest trees because these non-traditional sources of wood are so neglected, and are managed in many diverse ways.
- *Consider the multiple uses of biomass*, and the large quantities of biomass that may be available from different industrial processes such as sawmill waste, charcoal, etc.
- *Consider the different types of biomass resources* [e.g. woody biomass (firewood) and non-woody biomass (such as agricultural residues)].
- *The commercial value of biomass.* Bear in mind that biomass resources may be valued differently as an energy resource depending on the local, regional or national tradition and culture. For example, animal residues can play a significant role in some countries such India, but are hardly used in others.
- *Be aware of fuel switching.* As living standards increase, or people move to rural centres, fuel preferences may change.

- *Be aware of the increasing importance of secondary fuels*, e.g. biogas, ethanol, methanol, etc. (see also Chapter 4). These fuels are obtained from raw biomass and are used in increasing quantities in modern applications, both in the developed and developing countries.

Assessment methods for woody biomass

The method for estimating woody biomass whether directly or indirectly, is determined by a number of factors, including:

- its type,
- the area in which it is growing,
- its spatial pattern,
- its variability,
- its physical size.

Woody biomass is perhaps one of the most difficult but usually the most important measurements to make and hence its proper measurement is important. The following is a summary of the steps involved.

Multi-stage approach to woody biomass assessment

1 Review of existing data/maps for area
2 Low-spatial-resolution imagery, advanced very-high-resolution radiometer (AVHRR) (resolution 1–8 km)

- Large country or multi-country region
- Supported by limited truthing and existing biomass data
- Gives maps (scale 1:5,000,000) with broad vegetation types and rough biomass estimates

3 Spatial woodfuel-consumption patterns
4 Higher-spatial-resolution satellite imagery
5 LANDSAT TM and MSS (resolution 30–80 m)

- Small country or part of country
- Supported by ground truthing, aerial photos, airborne video plus existing data on biomass estimates
- Gives maps or overlays (1:250,000 to 1:1,000,000) with more reliable vegetation types and rough biomass estimates

6 Biomass reconnaissance inventory

- SPOT (Satellites pour l'Observation de la Terre; Earth-observing satellites) imagery and/or aerial photos
- Low-intensity inventory, perhaps with sub-plots for smaller vegetation

- Supported by cartographic or imagery maps and established regressions and correlations between measurable parameters and biomass
- Destructive sampling to establish regressions
- May include non-forest as well as forest areas

7   Biomass management inventory

- Higher intensity with sub-plots supported by maps in forest and non-forest areas

8   Conversion of industrial wood volumes to total above-ground biomass

- Supported by maps and/aerial photography, and established regressions and correlations between industrial wood volume and total biomass

Conversion of industrial wood volumes to total above-ground biomass

This data should be supported by maps and/aerial photography, and established regressions and correlations between industrial wood volume and total biomass.

Developing your assessment strategy

The following initial questions may be helpful in formulating your assessment strategy for woody biomass.

What types of woody vegetation are present in the area?

- Forests: planted or natural; main species
- Bush land
- Open woodland
- Trees in and around farming areas: woodlots, windrows, scattered trees in cropland, trees on compound
- Trees in public places: markets, roadsides, along canals

What is the condition of these vegetation types?

- Well maintained or neglected
- Gaps because of heavy cutting
- Natural regeneration
- Pruning, pollarding
- Collection of dead wood
- Fresh stumps
- Coppices
- Litter
- Erosion

Do you observe any transportation or trading of forest or tree products?

- Heaps of wood on the roadsides
- People transporting wood, charcoal, fruits, tree leaves, etc.
- People selling wood, charcoal, fruits, bark, roots, medicines, etc., in markets or elsewhere

Do you see any activity related to processing or utilization of tree products?

- Sawing or splitting
- Burning charcoal
- Fencing
- Building
- Processing of fruit
- Basket making
- Feeding leaves to cattle, etc.

Do you observe any activity related to tree regeneration and management?

- Tree nurseries
- Transportation
- Selling of seedlings
- Young tree or newly planted cuttings
- Pruning, clipping, thinning, clearing, coppicing

Measuring fuelwood resources and supplies

In estimating total wood resources and actual or potential wood supplies one must first make a clear distinction between standing stocks and resource flows; that is, the rate of wood growth, or yield. Other important distinction for energy assessment include competing uses of the wood, e.g. timber, construction poles, etc.

It is important to estimate the non-fuel uses for wood. Wood sold for purposes other than fuel has a much higher market value, and consequently normally has priority over use of wood as fuel. The branch wood and crooked stem wood, which might amount to 15–30% or more of the above-ground volume, may be left when the trees are felled, and could potentially be a source of fuel.

You should also take into account the considerable waste that is generated when trees are converted to wood products. Even the conversion of sawn wood into poles or timber produces more waste. All these waste materials are potentially burnable, especially if they are near to the demand, and must therefore be included in the assessment.

Where twigs and leaves are collected for fuel it is necessary to carry out destructive sampling of a small number of trees to provide measurements of the leaf, branch, and stem and root weights. You can then estimate the total available tree biomass per unit area:

- the fraction of the standing stock and yield that is actually accessible for exploitation due to physical, economic or environmental reasons,
- the fraction of the yield that can be cut on a sustainable basis,
- the fraction of the cut wood that is actually recovered.

Estimating stock inventories

The standing stock of trees is normally estimated by aerial surveys or satellite remote sensing to establish the areas of tree cover by categories, such as closed forest, plantations, etc. This information is then combined with the survey and mensuration data.

Note that estimates of tree stocks are always approximate. Most inventory data is for the volume of commercial timber, which are a small proportion of total standing biomass. The quantity of fuelwood biomass may greatly exceed the commercial timber volume.

Estimating supplies

The following example shows the estimated amount of wood that can be obtained from natural forest by (1) depleting the stock and (2) sustainable harvesting. Essentially, the method involves simple multiplication to adjust stock and yield quantities by the accessibility and loss factors. This model could apply equally well to a managed plantation or village woodlot (of course with different numbers), to estimating the effects of forest clearance for agriculture (partial or complete stock loss) and the evaluating the impact of fuel gathering on forest stocks. You will probably have to put more effort into measuring scattered non-forest trees because these important sources of wood are often neglected and involve considerable diversity in tree management (see Table 3.1).

The measurement of woody biomass

Techniques are available for measuring tree biomass by volume and by weight. For the reasons stated already in Chapter 2 weight is the most suitable measurement for fuelwood surveys. However, because the forest industry measures timber by volume, the techniques for determining the volume of tree stems and larger branches are far more developed. The techniques for assessing both weight and volume are outlined here. However, to enable comparison, all measurements should be converted to metric weight units (see also Appendix 3.2)

It is an easy matter to find the total weight of the tree, including the crown, if tables exist that give the relationship between stem and branch volume for the species being assessed. As there are standard techniques for calculating the volume of stems and crowns, it may be easier where tables are available to first calculate the volume and use this to estimate the weight.

Table 3.1 Example of a stock and yield estimation method: natural forest/plantation (hypothetical data)

		Assumptions	Stock data	Yield data
Supply factors				
A	Forest area	1,000 ha		
B	Stock density	200 m³/ha		
C	Stock volume	200,000 m³		
D	Mean increment			0.4 m³/ha/year
E	Sustainable yield			3.8 m³/ha/year
F	Gross sustainable yield (A × F)			3,800 m³/year
G	Fraction available for fuelwood		0.4	0.4
H	Fraction accessible		0.9	0.9
I	Harvest/cutting fraction		0.9	0.9
J	Gross sustainable harvest (G × I × J)			3,078 m³/year
K	Fuelwood sustainable harvest (K × H)			1,231 m³/year 1.23 m³/ha/year
Clear felling				
L	Gross harvest (C × I × J)			162,000 m³
M	Fuelwood harvest (M × H)			64,800 m³
N	Wet density (0.8 tons/m³)			
O	Net heating value (15 GJ/tonne or MJ/kg)			
P	Energy harvest: clear felling (N × O × P)			777 TJ
Q	Energy harvest: sustainable (L × O × P) GJ/ha/year			14.6 TJ/year 14.6
R	Other wood: clear felling, (M − N) × O			77,700 tonnes
S	Other wood: sustainable harvest, (K − L) × O			1,477 tonnes/year 1.47 tonnes/ha/year

TJ = terajoule = 1,000 GJ.

Further reading: Leach and Gowen (1987), pp. 93–94; Ramana and Bose (1997); Yamamoto and Yamaji (1997).

Satellite imaging and/or aerial photography are useful tools to estimate woody biomass. As branch volume and weight can vary from 10% for trees grown in uniform plantations to over 30% for free-standing trees it is essential to augment imaging data with field measurements. A literature search may also provide useful information. A combination of information from these sources should provide accurate estimates of biomass volume and weight estimates for individual sites.

Measuring fuelwood

Much fuelwood is collected by the head load. A number of head loads could be measured either by weight or volume and an average for a particular district or country should also be established. The head load size may differ considerably from district to district, region, etc. For example, in Tanzania the average of a head load was found to be about 26 kg whereas in Sri Lanka it was about 20 kg. Therefore size and weight must be established for each particular case.

Volume and weight

These two methods of measuring each have their drawbacks. If *volume* is used, then the conversion factor from the bundle to solid measure can vary enormously, depending on whether the head load consist of one large log or many small branches.

In some countries the *stere* or *stacked cubic metre* is the standard measure, but using this measure it is not possible to know the correct conversion factor to apply. If the stere is made up from bundles, then the conversion factor will be much lower than if it is made up from stacked stem wood; although the stacked measure is not an exact measure and can be up to 20% more than a true stere. This also applies to the other stacked cubit feet or the metric core. One advantage of the volume measure over the weight measure is that the volume of *wet* wood does not differ greatly from air-dry wood (up to approximately 5% difference); if a standard conversion factor to convert weight into volume is used, without accounting for the moisture content of the wood, there can be 100% difference in volume estimation, depending on whether the wood is green or oven dry.

It is important to note that in the case of domestic fuelwood the volume is not such a suitable measure to quantify. The wood is often of an irregular shape, and as the quantities used are relatively small it is usually much easier to determine the weight. *Weight* thus may be a much more convenient measure to use to ascertain solid volume, for the weight of a bundle of wood is easier and quicker to determine than trying to determine the gross volume of an irregular shaped head load of fuelwood. If the solid volume was to be measured then every piece of wood would have to be measured separately, or the water displacement method used. It is, however, important to know the moisture content of wood if weight is the measure for assessing the solid round wood volume. An additional problem with using weight as a measure to determine solid volume is that the weight depends on density, and the density within and between wood species is not uniform. If the moisture content is the same, *the energy given off from a piece of wood remains about the same on a weight basis irrespective of species.* In other words, the energy content per unit weight for different species of wood varies far less than the energy content per unit volume.

Projecting supply and demand

There are five main methods for predicting supply and demand: (1) constant-trend based projections, (2) projections with adjusted demand, (3) projections with increased supplies, (4) projections including agricultural land and (5) projections including farm trees. This is discussed in detail in Appendix 3.1 and in Box 3.2 which illustrates a hypothetical wood balance.

Box 3.2 Example of a constant trend–based projection: hypothetical wood balance

Table 3.2 Example of a constant trend-based projection: hypothetical wood balance

	1980	1985	1990	1995	2000	2005
Standing stocks, m³★	17,500	16,010	13,837	10,827	6,794	1, 520
Fuelwood yield, m³/year★	350	320	278	217	136	30
Consumption, m³/year★	600	696	806	935	1,084	1, 256
Deficit, m³/year★	250	376	529	718	948	1, 226
Population★	(1,000)	(1,159)	(1,344)	(1,558)	(1,806)	(2,094)

★10×3.

Assumptions: Fuelwood yield: 2% of standing stock (standing stock: 20 m³/ha). Population: 1 million in 1980, growth at 3% per year. Consumption: 0.6 m³/capita/year. Deficit is met by felling the standing stock.

Calculation method: Calculations are performed for each year (t, t+1, etc.), taking the stock at the start of the year and consumption and yield during the year:

$$\text{Consumption}(t) = \text{reduction in stock}\,(t, t+1) + \text{yield in year}\,(t)$$

(continued)

(continued)

$$\text{Stock}(t) - \text{stock}(t+1) + M/2 \times \left[\text{stock}(t) + \text{stock}(t+1) \right]$$

where M = yield/stock expressed as a fraction (0.02 in this case).
Hence to calculate the stock in each year:

$$\text{Stock}(t+1) \times (1 - M/2) = \text{stock}(t) \times (1 + M/2) - \text{consumption}(t)$$

Further reading: Leach and Gowen (1987), pp. 132–140; Openshaw (1998).

Forest mensuration

Mensuration – the measurement of length and mass over time – incorporates principles and practices perfected by land surveyors, foresters and cartographers. Forest mensuration is the tool that provides data on forest crops, or individual trees, felled timber and so on. The principles of measuring trees are given in several books (e.g. see Husch et al., 2003). However, there is only one satisfactory way to measure shrubs and hedges and that is to cut them down, weigh them and obtain a relationship between the volume of the whole shrub, including air space, and weight.

There are many techniques used by commercial forestry to measure forest, individual tree parameters, branches, bark, volume, weight, etc., that you can borrow to estimate total or partial biomass availability.

Individual trees should be described quantitatively by various measurements, or parameters, the commonest of which are:

- age,
- diameter of stem, over or under bark,
- cross-sectional area, calculated from diameter of stem,
- length or height,
- form or shape: trees are not cylindrical,
- taper or the rate of change of diameter with length,
- volume over or under bark; volume may be calculated to varying top diameters,
- crown width: a parameter that can be measured both in the field and/or aerial photographs, and
- wood density.

Measurements of tree crops, woodlands, plantations and forests require further measurements including:

- area: surveyed or estimated from maps or aerial photographs,
- crop structure: in terms of species, age and diameter,

- total basal area per hectare,
- total biomass, dry weight per hectare, and
- total energy resource per hectare.

For even-aged, uniform plantations of a single species the following measurements are also frequently used:

- average volume per tree,
- average stem basal area per tree,
- diameter of tree of average basal area, and
- average height, per tree (King et al., 1991).

Traditionally, foresters measure trees by stem volume or by weight, which includes the moisture content. It is possible to estimate the stem volume of a tree with height and diameter measurements, and the per hectare volume with average height and basal area measurements. The unit weight of these trees can be calculated if the density of the wood is known.

Field surveys providing on-the-ground biomass data have not been done, partly due to lack of interest in biomass, and partly because of the lack of appropriate calibration curves relating individual tree or bush biomass to easily measurable tree or bush dimensions. Extensive mensuration data is available for the very limited number of tree species commonly cultivated in plantations. When only a few species are present it may be possible to estimate the woody biomass of plantations using land use data combined with biomass tables that can provide the volume or weight of biomass for a given species. These tables are particularly useful where the vegetation is fairly uniform. At present very few biomass tables are available. Once drawn up, however, such tables allow the subsequent survey to be carried out rapidly.

Natural forests in subtropical environments are composed of a multitude of differing trees for which little or no information is available. The standing biomass of trees in natural tropical forests has been handled on a case-by-case basis (see Allen, 2004). Surveys have produced regression equations against stem diameter, stem circumference, stem basal area, tree crown dimensions and combinations of these on a case-by-case basis, resulting in many custom-made regressions, suiting the aims of the various researchers. For example, Tietema (1993) carried out studies in Botswana aimed at providing a set of regression equations relating external tree dimensions to total above-ground biomass. This was done by measuring the height and the diameter of the crown, together with the diameter of the stem at 'ankle height' (approximately 10 cm above ground level) of sample of trees. The trees were then cut down so that the weight of the total fresh biomass could be measured. For multi-stemmed trees the stem basal area and the tree weight were determined for the individual stems. The results show that in the regression of stem basal area against weight there is a great similarity between the regression lines of a range of trees species in Botswana and also in Kenya with three different species.

Thus, in general, the set of regression lines offers a realistic and flexible possibility of carrying out extensive on-the-ground surveys of tree standing biomass. This flexibility, according to Tietema, is very important in determining the effect of wood harvesting, establishing tree stock and the mean annual increment (MAI), and in assisting the interpretation of remote sensing data. However, it may not always be feasible to use destructive sampling as part of the survey, for example, when trees are incorporated in agricultural areas. (Further details are given in Appendix 3.2.)

The main techniques/methods for measuring woody biomass

This handbook considers in detail only the main techniques that can be applied to assess woody biomass for energy.

Determining the weight of trees

It is possible to determine the weight of specific tree species in natural environments using measurements of the stem diameter (breast height, 1.3 m, or 0.4 m), stem circumference, stem basal area (0.1 m), tree crown dimensions and combinations of the tree and crown dimensions. Due to the differing nature of the linear regressions obtained, making a comparison between species is very difficult. The example below is from destructive measurements for British native woodland species (Figure 3.1).

As mentioned above, commercial foresters are mostly interested in stem volumes, which give a considerable underestimate of the wood available for fuel. Branches and roots, which are important sources of fuel, make up well over 50% of the tree wood. Where wood is scarce, twigs and leaves are also collected for burning. When assessing biomass for energy purposes, your calculation should include leaves, twigs, branches, stems and roots. Note that tree roots can amount to some 30–40% of total woody biomass production, and approximately 55% of above-ground woody biomass production. However, unless there is a change of land use, or a severe shortage of fuel, tree roots are generally not burned, although this does not apply to roots of shrubs and bushes.

Destructive sampling of a small number of trees to provide sample measurements of the leaf, branch, stem and root volumes or weights will allow estimates of the available biomass per unit area to be made.

Once the regression for a species or group of woody biomass species has been established, the most useful and easiest-to-measure regression is probably that between stem basal area and tree weight. This technique requires that only two measurements of the diameter for each stem (single or multiple stemmed trees) are required. The technique is straightforward and requires only simple equipment: callipers, scales and drying apparatus for establishing the regression. Thereafter only callipers are required for rapid measurement of the necessary

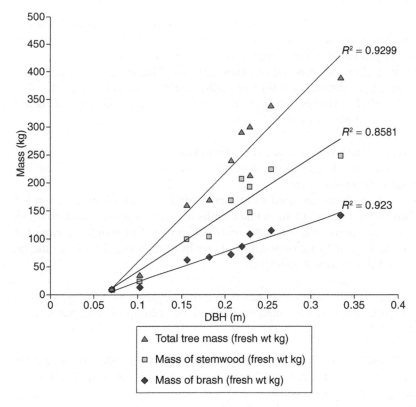

Figure 3.1 Example of destructive measurement (British woodland species). DBH, diameter
   at breast height.

sample of trees. Establishing the best sampling technique for small or large areas
is probably as important as the actual measurements themselves.

   To obtain weight estimates, a sample of trees has to be felled to determine
the green, air-dry and possibly oven-dry density (weight/unit volume). While
wood loses weight when it dries, there is little shrinkage until wood gets down
to about 10% moisture content, so knowing volume and average density, the
weight (air-dry or oven-dry) can be determined with sufficient accuracy.

Techniques for measuring the volume of trees

Traditionally, to measure the volume of trees, foresters use:

* diameter at breast height (DBH),
* total height and crown measurements (diameter plus depth) to estimate
   individual tree volumes, and
* mean height, basal area at breast height and mean crown measurements.

Volume measurement of trees and roundwood

The volumes of trees and roundwood are commonly measured both for the purposes of management and for their sale. By roundwood is meant the products that are produced when a tree is crosscut into lengths. These products may be sawlogs, pulpwood, fencing or other material. The unit of volume is the cubic metre. Calculating the volume usually requires a measurement of diameter and length, but may be measured in different ways according to the nature of the wood you are measuring.

- *Diameter.* Diameter is measured in centimetres, rounded down to the nearest centimetre. For example diameters measured as 14.9 or 14.4 cm are rounded down to 14 cm.
- *Length.* Length is measured in metres. Lengths under 10 m are rounded down to the nearest 0.1 m, while lengths over 10 m are rounded down to the nearest metre. Thus lengths of 16.75 and 18.3 m would be rounded down to 16 and 18 m, respectively. Lengths of 6.39 and 3.95 m would be rounded down to 6.3 and 3.9 m.

When measuring the length of a felled tree the length is usually measured to a top diameter of 7 cm. The different methods of volume measurement are as follows.

- *The 'length and mid diameter' method*: this is used for felled trees and for situations where it is reasonably easy to get to the point at which the mid diameter has to be measured.
- *The top diameter and length method for sawlogs*: special conventions apply where reasonably uniform logs are sold in quantities of at least a lorry load. Their volume can be estimated from top diameter and length making an assumption for taper.
- *The top diameter and length method, for small roundwood*: volume may also be measured from top diameter, again making an assumption on taper.
- *Mid diameter and length volume measurement method*: the volume is dependent on the length of the tree or roundwood piece and the diameter at the middle of the rounded down length. Length and diameter are of course measured according to the conventions stated above. Examples of the mid diameter and length method of measuring volume are given in Box 3.3.

Box 3.3 Examples of the mid diameter and length method of measuring volume

Roundwood less than 20 m in length. The measured length of the tree is 4.45 m, rounded down to 4.4 m. The measured mid diameter (at 2.2 m) is 14.6 cm, rounded down to 14 cm. From the tables, the volume for a length of 4.4 m and 14 cm diameter is 0.068 m^3.

Roundwood greater than 20 m in length. If the length of the tree to 7 cm top diameter is greater than 20 m then the convention is to measure the volume in two lengths (by dividing the total length of the tree by two). First, measure the length of the bottom (butt) half and round it down. The second length will be the remaining length. For example:

Tree length = 37 m, which divided by two is 18.5 m

$\text{First}\left(\text{butt}\right)\text{length} = 18\text{ m}\left(\text{mean diameter taken at }9\text{ m}\right)$

$\text{Second length} = 19\text{ m}\left(\text{mean diameter taken }9.5\text{ m along the length}\right)$

The volume of each length is then obtained from the mid diameter/ length table, and the totalled volumes is the volume of the tree stem. (See www.woodlander.co.uk/woodland/.)

If local volume tables are available for the tree species being assessed it is possible to use the diameter alone to determine both the volume of the stem and total tree volume for single species stands. On standing trees the most common diameter measured is the reference diameter, usually measured at 1.3 m above the ground level, and commonly known as the DBH. The DBH is important because it is relatively easy to measure directly, and from which the cross-sectional area and the volume can be computed. DBH is widely used as an independent variable in estimating total wood weight in a tree, but this is applicable only to well-defined species and mostly to plantation and commercial forestry. However, even within the same species there can be a substantial variation in volume for trees of the same DBH. Therefore, general volume tables should be used that take height into consideration as well. Appendix 3.2 provides further details of how to measure tree volume.

Foresters have devised formulae to calculate weight from volume, so that it is an easy matter to find the total weight of the tree, including the crown, from tables that give the relationship between stem and branch volume. How useful these techniques can be for measuring fuelwood is not always clear. Furthermore, there are also standard techniques for calculating the volume of stems and crowns. It may therefore be easier to first calculate the volume and use this to estimate the weight.

If the average height and the tree taper are known then, for a particular species, it is possible to estimate both the volume of the stem and the total volume of the tree above the ground.

The simplest way to calculate the volume of a tree is by using the formula:

$$v = \pi r^2 h \times f$$

where v is the volume of the tree, r is the radius at breast height, h is the total height and f is the reduction factor to allow for the taper on the tree.

The reduction factor (f) can range from 0.3 to 0.7 and has to be calculated from felling a number of trees and measuring individual logs.

Similarly, it is also possible to estimate the crown wood volume from the stem volume. A number of trees are felled to obtain the ratio of wood in the stem to that in the crown. This calculation is then used to estimate crown volume in the remaining trees from stem volume alone.

In closed stands, the per hectare stem crop volume (V) is given by the formula:

$$V = G \times H \times F$$

where G is the mean basal area per hectare, H is the mean height and F is the mean reduction factor.

The basal area (G) is easily determined using an angle count measure (or relascope technique). A hypsometer can be used to obtain the height of a number of trees to determine mean height (H). The principles of measuring trees are given in several books (e.g. Philip, 1983; Husch et al., 2003). See Box 3.4.

Box 3.4 Compilation of branch volume

Step 1   Calculate the volume of each branch section using the formula:

$$\text{Volume} = \frac{(A^1 + A^2)L}{2}$$

where

A^1 = cross-sectional area (outside bark, o.b.) of the bottom end of the branch section,

A^2 = cross-sectional area (o.b.) of the top end of the branch section,

L = length of the branch section.

Step 2   Sum the volumes per tree.
Step 3   List the trees by species, DBH class and branch volume.
Step 4   Perform a multiple regression analysis in order to find the best fit equation for the relationship between DBH class and branch volume.
Step 5   Tabulate the equation results.
Step 6   Prepare a branch volume summary for each stratum by multiplying the number of trees per hectare (from the stand table) by the branch volume (from the table prepared in step 5) for each DBH class.

Compilation of tree volume

This process consists of calculating the volume of the tree stem bole from the individual section measurements taken during the volume and defect study. Individual tree bole volumes are used in the compilation of the sample plot volumes, calculated on a species basis.

Volume of felled trees

Step 1 Calculate the gross volume of each section of the tree stem bole using the simple formula:

$$\text{Volume} = \frac{(A^1 + A^2)L}{2}$$

where

A^1 = cross-sectional area (inside bark, i.b.) of the bottom of the section,

A^2 = cross-sectional area (i.b.) of the top end of the section,

L = length of the section.

Step 2 Sum the section volumes per tree.
Step 3 List tree data by species showing diameter, height and volume.

Further reading: King et al. (1991), pp. 11.1 ff; Ashfaque (2001); Openshaw (1998).

Allometric methods

Allometric, or dimensional, analysis is an old and widely used technique for estimating timber volume and weight for commercial species. A number of trees are felled, and the various parts (branches, leaves, twigs, etc.) are weighed, and then related by a regression analysis to an easily measurable variable on the standing tree, such as DBH. It is then possible to estimate the weight or the volume using stand tables, which tabulate the diameters of tree stems found in a unit area. Using the regression formula calculated from the sampling it is then possible to calculate the weight (or volume) of biomass for each tree using its stem diameter, and thereby to estimate the biomass for the whole stand.

This technique works well for managed plantation. However, it is not suitable where stands of trees are composed a variety of trees with different shapes and densities, or where trees have been extensively managed by pruning, lopping, etc. (e.g. farm trees). Thus, there is no single universal formula for measurement applicable to all trees. Conifers differ from hardwoods, tropical woodland (savannahs) trees from tropical forest trees and so on. Methodologies should be chosen according to the particular situations and circumstances.

For forest species, especially tropical forest species, selected trees are usually measured along line plots at specific intervals using such measures as diameter (DBH), height and crown measurements. Trees outside the forest are measured in a similar fashion. However, these measurements tell us nothing about the area covered by the different woody types. For dense formations, area data can be obtained from satellite pictures, and then estimates of biomass volume and weight can be made for individual sites.

Satellite photography has great potential as it offers the possibility of covering large areas with a high degree of definition. But the technique still needs improvement. If the trees are scattered and sufficient resolution is possible the estimated for height and crown dimensions and even perhaps DBH of the tree can be estimated. If the woodlands are relatively undisturbed, and contain a cross-section of most ages and stem diameters, the approximate volumes and/or weight may be estimated from crown cover. Of course, some field work is necessary to obtain this volume relationship. Crown cover may also be estimated directly from ground measures and from aerial observations (see Chapter 8).

Measuring growing stock of trees for fuel

The section above gives a measure of growing stock, but increment or annual yield is required as well so that you can estimate the potential sustainable harvest for fuelwood. If it is possible to calculate the total growing stock it is then possible to estimate the annual increment. Note that you cannot assess annual increment by simply dividing growing stock of fuelwood plantations by the rotation age, because trees do not put on equal annual quantities of biomass throughout their lifetime.

Some fuelwood crops may have rotations as short as 1 or 2 years, whereas in natural forest the rotation may be in excess of 100 years. Natural forest has a very large growing stock, but a relatively low annual yield, of the order of $2-7$ m³/ha. Figure 3.2 shows the relationship between age and growing stocks based on data from British woodlands. Better management can reduce the rotation and increase the annual yield so that there is a quicker turnover of stock. You should also make a thorough literature search of work on whole tree measurements and biomass grown on various sites and under different rainfall regimes (Hemstock, 1999).

Measuring annual yield of trees

Biomass production depends on the quality of the growing site, rainfall, tree species, planting density, rotation age and management techniques. Wider spacing produces larger trees, and planting more trees per hectare will not increase production once there is canopy closure. However, biomass production per unit area is more or less constant for trees of the same age on similar sites for a

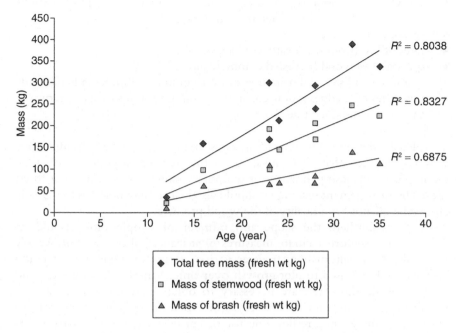

Figure 3.2 Relationship between mass and age of trees in British woodlands.

fairly wide range of planting densities. The current annual increment (CAI) is the increase in volume over a year. The MAI is the average CAI over a stated number of years. Note that the CAI will be high when the trees are young and growing, but may be zero in a mature forest where the canopy is closed.

While satellite photography may be valuable in determining areas of dense woody biomass, it cannot provide accurate data on either growing stock or annual increment. There is, at present, no substitute for ground measurements.

For plantations, you can calculate the CAI from the total growing stock. The CAI is estimated by dividing the growing stock by half the rotation; that is:

$$I = GS \times R/2$$

where I is annual increment, GS is growing stock and R is rotation age. This formula works well except where the rotation age is small (for crops with a 1 year rotation $GS = 1$ and not $2GS$). Note that you cannot assess annual increment simply by dividing growing stock by the rotation age, because trees do not put on equal annual quantities of biomass throughout their lifetime.

Yield of wood per unit area

The next step is to determine the weight or volume of wood per unit area. To calculate the sustained yields of wood and residues it is necessary to know:

- growth rates: foresters measure volume increments by taking periodic measurements of stem diameter, height, etc.,
- reproductive rates,
- rates of loss (from tree death and harvesting),
- quantities of wood harvested commercially,
- quantities and spatial distribution of non-commercial fuelwood collected,
- quantities of residues produced, and the amount it is necessary to return to the soil for sustained tree growth.

To calculate the annual increment, it is best to measure trees at periodic intervals, making sure that all removals from the trees are accounted for. One method is to establish permanent or temporary sample plots in fields or forest sites. The management techniques should also be carefully noted as this greatly affects the quantity and quality of the wood produced.

Once established, the scope and number of sample plots should be enlarged to get more accurate measures of at least all above-ground woody materials. As an alternative, trees at various stages of development could be measured and used to plot growth over time. Sample plots for trees and shrubs outside the forest are necessary to obtain better removal statistics from this important source of supply. For this technique, it is important to ensure that the agency responsible for the project has the necessary infrastructure to carry out this task.

Alternatively, yield tables are available that predict the growth and yearly yield of single species stands. The rotation or lifetime is defined as the point of maximum MAI. Yield tables and volume tables are usually available for single-plantation species for stem volume only, but can be adapted to give total volume by determining the relationship between stem wood and total wood.

Measuring tree height

This is important since it is often one of the few variables used in the estimation of tree volume and thus it is a common attribute measured in forest inventories. 'Tree height' can have several meanings and can lead to practical problems (see Appendix 10.1, Glossary, under Tree height measurement, for definitions).

To be able to measure the height of a tree, the top must be visible from a point from which it is feasible to see most of the stem bole. The methods used generally depend on the size of the tree to be measured. Direct methods (rarely used) involve climbing or using height-measuring rods, whereas indirect methods (commonly used) use geometric and trigonometric principles using a theodolite or clinometer (see Figure 3.3).

Measuring tree bark

Bark is the outer sheath of the tree. Some tree species shed bark while others have persistent bark that is removed when trees are cut. Knowing bark volume

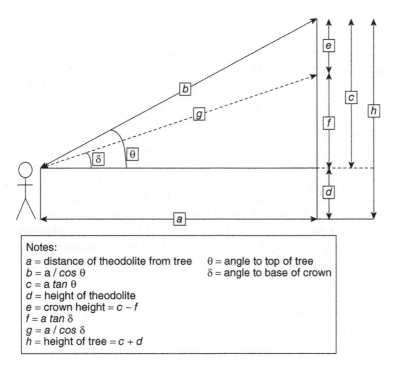

Notes:
a = distance of theodolite from tree θ = angle to top of tree
$b = a / cos\ \theta$ δ = angle to base of crown
$c = a\ tan\ \theta$
d = height of theodolite
e = crown height $= c - f$
$f = a\ tan\ \delta$
$g = a / cos\ \delta$
h = height of tree $= c + d$

Figure 3.3 Height measurement and crown dimensions.

is important in commercial logging when it is a valuable saleable product, or when it is collected as fuelwood. On felled trees bark thickness is measured directly at the cut ends.

To measure bark thickness:

- place the bark gauge perpendicularly against the tree and push the arm until the whole of the bark has been traversed,
- read the bark thickness on the scale,
- take another measurement at a point diametrically opposite the first point of measurement,
- take the arithmetic mean.

Methodology for assessing energy potential from forestry/crop plantations

Despite the expected increased role of dedicated forestry/crops energy plantations, currently there is little commercial experience with such plantations. The only country that has large-scale plantations is Brazil where there are some 2.5 Mha of eucalyptus used in the industrial production of charcoal for steel, metallurgy, cement, etc. Following is a brief summary of the methodology for assessing energy potential from dedicated plantations. For further details consult Kartha et al. (2005), pp. 104–118.

Assessing the potential resources from dedicated energy plantations should be much simpler than residues and forests (natural, on-farm, commercial plantations for non-energy use, etc.) for various reasons. First, there are fewer parameters to consider (i.e. energy is the main purpose rather than one of many, as is often the case when energy is just a by-product). You are basically concerned with harvested area, yield per hectare, energy content, etc., using conventional forestry or agricultural criteria. Second, dedicated energy plantations respond to specific parameters required in modern applications such clear financial criteria, which is widely understood (i.e. is there a market and, if so, is the cost competitive, or can the land be put to better alternative uses?).

The methodology for assessing dedicated energy plantations can be used at any scale (village, regional, national) and at any desired level of disaggregation and type of energy crop. Kartha et al. (2005) have identified six major steps for wood energy, which are summarized in Box 3.5 below.

Box 3.5 Major steps for assessing wood energy from plantations

Step 1   Estimate areas of potentially available and suitable land

Land is the key factor and thus you need to classify land into various categories: land availability and land suitability for energy plantations. This is not as simple as it may look given the multiple uses of land, particularly when it comes to food production, which should be given a higher priority.
    Land not available includes:

- legally protected land (e.g. national parks, forest),
- socially protected land (e.g. amenity areas, forests, woodlands),
- built-over land (e.g. urban, industrial areas).

Land not suitable:

- climatic constraints (e.g. low precipitation, high temperature),
- terrain and soil constraints (e.g. steep slops, rocky soil),
- remoteness and lack of infrastructure (e.g. difficult access, weak local energy demand),
- productive land under high-value crops (e.g. high-value timber plantations).

Step 2   Estimate yields associated with areas of suitable land

After assessing land availability and suitability, the next step is to find the yields for which you can use standard production practices. Most of the agricultural and forestry production models can be used to estimate yields.

Step 3   Calculate the gross potential resource (by land type and/or sub-region)

This step merely involves multiplying the area of available and suitable land (step 1) by associated yields (step 2) to produce an initial estimate of the gross potential resource from the energy crop in question. This is required as both a total as well as by sub-region or land type, depending on the level of land disaggregation used in step 1 (see Kartha et al., 2005).

Step 4   Estimate production costs and delivered energy prices

The costs of growing biomass for energy are similar to those of agriculture and forestry, although with some differences. Using the terminology of forestry (Kartha et al., 2005) the basic cost categories are:

- the *stumpage* cost (the capital cost of establishing and maintaining the crop until it is harvested),
- harvesting and haulage to the fuel-preparation site,
- fuel preparation (e.g. chipping and drying),
- transportation (to the market or power plant),
- overheads and fixed costs.

Step 5   Estimate delivered energy prices (and ranges)

Once you have the cost of all items identified in the previous steps, it is not difficult to calculate the cost of energy generated and associated sale prices. The best way is to use a computer spreadsheet to tabulate all costs and revenue for each year (e.g. number of harvest rotations), discount these at chosen interest rate(s) and calculate measurers of financial worth for hypothetical plantation projects [e.g. benefit/cost ratio, net present value (NPV) and internal rate of return (IRR)].

Step 6   Final integration and assessment of energy crop potential

The standardized results from step 5 can be used in three main ways to arrive at the final analysis of the total energy potential and associated prices.

1   For each discount rate and assumed net margin, locations can be filtered out as being suitable for energy crops. Rejections can be based on very high break-even costs relative to competing energy prices.
2   For each discount rate and assumed net margin, the same process can be used to tabulate and plots energy outputs against price to produce a cost/supply curve.
3   If the data is available on how crop yields respond to site quality and greater production inputs and costs, see Kartha et al. (2005).

Agro-industrial plantations

These crops are extremely diverse and no single method is suitable for determining the standing volume or weight and yield for all crops. First, obtain the information already available by carrying out a literature search. A field survey may then be necessary.

On-farm trees

Unfortunately little reliable data exists concerning on-farm trees, even though they are extremely important as a fuelwood source. A field survey is the only way to collect accurate, detailed data for on-farm trees. Farm trees are managed in many ways, and some diameter classes and ages are usually absent in any one sample. If specific sample plots are chosen, then average volume estimates for above-ground biomass per unit area can be determined. If the density of the wood from the particular tree species is known, then the weight of this volume can always be estimated.

For example, one approach tried in Kenya was to assess the standing volume by measuring individual trees in sample areas. Yield was estimated from notional rotations of the various tree species. These measurements excluded shrubs. It is possible to estimate increment and yield for individual species if growth rates, soil types and rainfall are known. This will require a detailed ecological survey. The practical measurement of increment and yield of the total standing biomass is only possible by taking regular measurements (at least once a year, ideally from special inventory plots established in villages).

The next step is to collect information on wood consumption (see Chapter 5). This information is best obtained from surveys and questionnaires directed to the farmer and or the householder. The type of questions asked should include:

- how individual farmers manage their trees,
- the exact location of the tree, and the parts from which the biomass are collected,
- who has access rights to the biomass from these trees.

With the yield and consumption data collected, you can then create time-series data for production and consumption of biomass.

Hedges, shrubs and bushes

Hedges, shrubs and bushes are favourite collecting sites for woody biomass fuel, but standing measurements of such biomass is difficult. One solution is to measure height and crown and then cut and weigh a sample of shrub plants, together with the dead branches and twigs that collect underneath.

Processed woody biomass

Besides measuring the growing stock and annual yield of woody biomass and the annual production of plant and animal residues, it may be important to try and measure the supply of modified or processed biomass. Modification produces a waste product that can be burned, as in the de-husking of cereals and coffee, the conversion of saw logs into sawn wood or the preparation of copra from coconuts (see also Chapter 4).

Sawmill waste

Sawmill waste is a very important source of fuelwood as large quantities of waste can be generated. Sawmill waste can be divided into slabs or off-cuts, bark (if separated from the slabs) and sawdust. The amount of waste produced from a saw log depends on:

- the diameter of the log to be cut,
- the sawing method,
- the market for the sawn wood.

A frame saw is often slightly more efficient than a band saw, and both are more efficient than circular saws. Also, a market for small pieces of sawn wood will decrease the amount of waste. Much depends also on the current market value of wood.

The quantity of biomass available after processing is therefore calculated from the input of raw material. The use to which it is put then depends on the market. For example, the slabs and off-cuts may be consumed as a boiler fuel to drive the mill, or they may be used as the raw material for fibre board or particle board production, making pellets, etc.

To determine uses or potential uses for this biomass it is necessary to carry out a sample survey at sawmills, board factories, pulp mills and other wood-consuming industries, classified by site, type and location. Sample surveys could be undertaken to determine the amount of such waste, and its use.

Charcoal

Charcoal, by far the most important processed biomass fuel, is produced by burning biomass, usually wood, in a restricted supply of air. An estimated 200–300 million people use charcoal as their main source of energy around the world, and it is an important urban household fuel in many developing countries. African countries consume over 24 Mt of charcoal annually, which is more than half the entire global production. Production of charcoal is expected to increase significantly; particularly in Africa. This trend is predicted to continue in line with population increases. Even so, these estimates may be very conservative. Accurate data is difficult to estimate because charcoal production is an activity that is an integral part of the informal economy of many rural communities.

Charcoal production is a primary or secondary activity for many millions of rural labourers in developing countries, and is one of the rural activities that brings in cash. Many charcoal makers are itinerants. As such charcoal production may be semi-legal or illegal. It is important to record the legal status of charcoal making, as charcoal made illegally is often made using inefficient technology that has remained the same for centuries. One of the reasons is that charcoal is mostly an activity of the poor who are struggling to survive, let alone invest in technological improvements. The low energy efficiency (e.g. 12% in Zambia, 11–19% in Tanzania, 9–12% in Kenya) results in considerable waste of wood resources as well as having serious environmental impacts (recent data indicate that emissions are much higher than previously thought). There are very few countries that use charcoal for industrial purposes, such as Brazil, which might have put some resources into research and development, albeit on a very limited scale.

There is greater control over the process if retorts or brick or steel kilns rather that earth kilns are used. To obtain high yields of good-quality charcoal, the operator of an earth or pit kiln has to be skilful and alert. But despite the relative inefficiency of earth kilns, they are generally the most appropriate technology for the woodland, savannahs and rangeland areas where most charcoal is produced. Here the producer moves from site to site. Only when there is a concentrated amount of biomass within a small radius should a producer think of capital-intensive technology. Even with earth kilns, though, the efficiency can be improved by training and adopting certain techniques such as drying the raw material and constructing the kiln properly.

Measuring charcoal

Surveys should therefore be undertaken to record the chain of activities and often very large number of players from tree to market. This should include production sites, the technology employed, the skill of the people involved in the production, moisture content of the raw materials, chemical properties of the wood and the skill of the operator, all of which will greatly influence the quality and quantity of the charcoal. You should also record regulations, methods of transportation and the overall economics.

The following steps should be followed to estimate charcoal production and the raw wood material required.

- Identify a good stratified sample of charcoal producers.
- Record the type of kiln used by each producer.
- Estimate the quantity of wood (either volume or mass) and its moisture content utilized by each producer. This can then be checked against growing stock and yield figures from the particular tree formations.
- Estimate the quantity of charcoal produced by each producer. This will depend on the moisture content and the density of the wood used to make it, the equipment used to make the charcoal and the skill of the producer.

Note that if yield is measured on a wet or air-dry (as opposed to oven-dry) basis (weight of charcoal output divided by wet weight of wood input), water increases the weight of wet wood in the denominator of the charcoal yield equation.

- Record the quantity of each type of charcoal produced – lump charcoal, powder and fines – and any other marketable products such as creosote. The bulk of the fines produced is usually left at site, although it is possible to collect it if there is a market. Up to 25% of the charcoal by weight could be in the form of powder and fines. If it is made from such materials as coffee husks, all the charcoal produced is either powder or fines.
- Record the uses to which the charcoal is put. The bulk of fines are usually left on site, although in some cases it is injected into a charcoal blast furnace, or collected up and sold if there is a market.
- Record the production cycle: quantity of charcoal produced and time taken.
- Check wood consumption against growing stock and yield figures from the trees providing raw wood material.
- Calculate the yield obtained for each producer.

A separate column in the energy-accounting system should be reserved for wood or charcoal, because, as mentioned above, the production technology entails a considerable loss of raw material that should be accounted for. If desired, you can convert the quantity of charcoal back into its wood equivalent (see Box 3.6).

Knowing the input of the raw material and the output of saleable charcoal, you can calculate a conversion factor. Charcoal is such an important fuel that it is essential that the conversion factors are as accurate as possible, especially when calculating woodfuel use in all its forms for energy. Conversion factors differ according to the type of technology and varying moisture contents, as stated previously (Table 3.3). However, there is usually some waste produced

Table 3.3 Conversion factors per tonne of charcoal sold. Average volume is 1.4 cm/t at 15% mc db.* Moisture contents are shown on a dry basis. Unit is a cubic metre of roundwood.

Kiln type	Conversion factor (%)					
	15% mc	*20% mc*	*40% mc*	*60% mc*	*80% mc*	*100% mc*
Earth kiln	10	13	16	21	24	27
Portable steel kiln	6	7	9	13	15	16
Brick kiln	6	6	7	10	11	12
Retort	4.5	4.5	5	7	8	9

It is assumed that the fines are briquetted in the retort.

*With softwood about 60% more volume is required per tonne of charcoal; with dense hardwoods such as mangrove species about 30% less volume is required.

between the producer and the marketplace, especially as it may be transported some distance: 100 km or more is common. Some lump charcoal will powder and be lost or settle at the bottom of the bag, and not be of much use for burning. Transport can also be a critical component in the accessibility factor. Taking these considerations into account, a conversion factor of 12 m³ of wood raw material (about 8.5 t) per tonne of charcoal is probably more accurate than the FAO standard adopted by the United Nations of 6 m³ per tonne of charcoal (see Box 3.6).

Box 3.6 Converting charcoal to roundwood equivalent

Three major problems arise if charcoal is converted back to roundwood equivalent: wood density, moisture content of the wood and conversion method, all of which need to be known before conversion can be carried out. The density of the wood governs the yield of charcoal and thus a given volume of charcoal will give different weights of charcoal, depending on the species, moisture content, technology, etc. Moisture content also has an important effect on the yield of charcoal; as noted above the drier the wood the greater the yield of charcoal. The method used to produce charcoal can also affect the yield considerably. For example, the range for average tropical hardwoods at 15% moisture content can be from about 4.5 m³/tonne in a metal retort to about 27 m³/tonne at a 100% moisture content in a poorly designed kiln. Most of the charcoal produced in most developing countries is produced in earth kilns whose conversion factor can vary from about 10 m³/tonne of charcoal up to 27 m³ (see Openshaw, 1983).

Further reading: Openshaw (1983); Bialy (1986); Emrich (1985); Hollingdale et al. (1991).

The properties of charcoal depend not only on the moisture content of the wood, the type of wood and its chemical composition, but also significantly on the carbonization temperature. Carbonization at a low temperature produces a charcoal with a high level of volatiles known as 'soft and black' charcoal, which is mainly consumed in the domestic market. High-grade or 'white' is produced at very high temperatures, and is used as a reducing agent in the iron-making industry.

Samples of charcoal should be tested for carbon, moisture and ash contents, and energy value (see Box 3.7). It is impossible to obtain more than 50% conversion by weight of wood to charcoal. In practice, the upper limit is about 30–35%. If charcoal is carbonized properly it consists of at least 75% carbon by weight, and has an energy value of approximately 32 MJ/kg, compared to around 20 MJ/kg for dry wood. Some 'charcoal' may turn out to be little more than charred wood with a carbon content just over 50%. The energy value will then only be slightly more than dried wood. The charcoal yields from non–woody plants are slightly lower, as they only contain between 45 and 47% carbon (see Box 3.7).

Box 3.7 Analysing the composition of charcoal

A few methods have been developed to analyse the raw materials and products of the charcoal-making process (e.g. sample preparation techniques, testing of physical properties and chemical analysis) details of which can be found in Hollingdale et al. (1991) pp. 93ff. Chemical analysis is particularly important. For example, to find the gross calorific value (GCV) of a charcoal, a known quantity is burned under strictly controlled conditions in oxygen to ensure complete conversion to charcoal to its combustion products. The heat released by this combustion is determined on the basis of the following equation:

$$\text{Heat release} = \text{mass of apparatus} \times \text{specific capacity}$$
$$\text{of the apparatus} \times \text{temperature rise}$$

Equally, the moisture content of a charcoal sample (this represents the water that is physically bound in it) can be found by driving off free moisture from a sample in an oven and recording the mass loss.

Charcoal is usually sold by volume – per standard bag or basket, per tin or per pile – but sometimes directly by weight. Most frequently charcoal is sold by the standard bag, which can vary from area to area and country to country. The weight of charcoal depends on the moisture content and on the density of the parent wood assuming it has been completely or near completely carbonized.

It is important therefore to know the species from which charcoal is made. For example, charcoal made from tropical hardwoods (with a volume of about 1.4 m³ and 15% mc) will weigh approximately 33 kg per bag whereas charcoal made from softwood will weigh, on average, about 23 kg per bag.

When charcoal is sold by the tin, this can vary in size. For example, a 20 l paraffin tin will contain about 7 kg of charcoal if tropical hardwood is used. The tin and also the bag are sold at prices that fluctuate according to season, inflation, etc., while a pile is usually sold at a fixed price and therefore the quantity of the pile varies from season to season and over time.

Charcoal efficiency can be defined either in terms of *weight* or *energy*.

$$\text{Weight} = \frac{\text{Charcoal output}\,(\text{kg})}{\text{Wood input}\,(\text{kg})}$$

(continued)

(continued)

$$\text{Energy} = \frac{\text{Charcoal output}(\text{MJ})}{\text{Wood input}(\text{MJ})}$$

It is important to note that the heating value of the primary end product (charcoal in this case) is determined by its carbon content. The formula for the relationship between carbon content (C) and HHV (dry basis) of combustible fuels can be described as follows:

$$\text{HHV} = 0.437 \times C - 0.306 \, (\text{MJ/kg})$$

Appendix 3.3 provides a short example of charcoal production and use in small islands: A case study of Aassiriki Village, Moso Island, and Vanuatu.

References and further reading

ADLAR, P.C. (1990) *Procedures for monitoring tree growth and site change*. Tropical Forestry Papers no. 23. Oxford Forestry Institute, Oxford.

ALDER, D. (1980) *Forest volume estimation and yield prediction*, vol. 2. FAO Forestry Paper no. 22/2. FAO, Rome.

ALLEN, A.B. (2004) *A permanent plot method for monitoring changes in indigenous forests: a field manual*. Christchurch, New Zealand.

ASHFAQUE, R.M. (2001) General position paper on national energy demand/supply. In *Woodfuel production and marketing in Pakistan*. Sindh, Pakistan, 20–22 October. RWEDP/FAO Report no. 55. Bangkok, pp. 17–36.

BIALY, J. (1979) *Measurement of energy released in the combustion of fuels*. School of Engineering Sciences, Edinburgh University, Edinburgh.

BIALY, J. (1986) *A new approach to domestic fuelwood conservation: guidelines for research*. FAO, Rome.

BLYTH, J., EVANS, J., MUTCH, W.E.S. AND SIDWELL, C. (1987) *Farm woodland management*. Farming Press, Ipswich.

EMRICH, W. (1985) *Handbook of charcoal making*. D. Reidel Publishing, the Netherlands.

ETC Foundation (1990) *Biomass assessment in Africa*. ETC (UK) and World Bank.

FORD, E.D. AND NEWBOULD, P.J. (1970) Stand structure and dry weight production through the Sweet Chestnut (*Castanea sativa* Mill) coppice cycle. *Journal of Ecology* 58, 275–296.

FORESTRY COMMISSION (1984) *Silviculture of broadleaved woodland*. Forestry Commission Bulletin 62. HMSO, London.

FORESTRY COMMISSION (1988) *Farm woodland planning*. Forestry Commission Bulletin 80. HMSO, London.

FORESTRY COMMISSION (1998) *Sample plot summary data, computer printout*. Mensuration Branch, Alice Holt Lodge, Surrey (unpublished).

HEMSTOCK, S.L. (1999) *Multidimensional modelling of biomass energy flows*. PhD thesis, University of London.

HEMSTOCK, S.L. AND HALL, D.O. (1995) Biomass energy flows in Zimbabwe. *Biomass and Bioenergy* 8, 151–173.

HOLLINGDALE, A.C., KRISNAM, R. AND ROBINSON, A.P. (1991) *Charcoal production: a handbook*. CSC, 91 ENP-27, Technical Paper 268. Natural Resources Council and Commonwealth Science Council (CSC), London

HUSCH, B., BEERS T.W. AND KERSHAW, J.A (2003) *Forest mensuration, measurement of forest resources book*, 4th edn. John Wiley, Chichester.

IEA/OECD (1997) *Biomass energy: key issues and priorities needs*. Conference Proceedings. IEA/OECD, Paris.

IEA/OECD (1998) *Biomass energy: data, analysis and trends*. Conference Proceedings. IEA/OECD, Paris.

KARTHA, S., LEACH, G. AND RJAN. S.C. (2005) *Advancing bioenergy for sustainable development: guidelines for policymakers and investors*. Energy Sector Management Assistance Programme (ESMAP) Report 300/05. The World Bank, Washington DC.

KING, G., MARCOTTE, M. AND TASISSA, G. (1991) *Woody biomass inventory and strategic planning project (draft training manual)*. Poulintheriault Klockner Stadtler Hurter, p. 11ff.

LEACH, G. AND GOWEN, M. (1987) *Household energy handbook: an interim guide and reference manual*. World Bank Technical Paper no. 67. The World Bank, Washington DC.

OPENSHAW, K. (1983) Measuring fuelwood and charcoal. In *Wood Fuel Surveys*. FAO, Rome, pp. 173–178.

OPENSHAW, K. (1998) Estimating biomass supply: focus on Africa. In *Proceed. Biomass energy: data analysis and trends*. IEA/OECD, Paris, pp. 241–254.

OPENSHAW, K. (2000) Wood energy education: an eclectic viewpoint. *Wood Energy News* 16(1), 18–20.

OPENSHAW, K. AND FEINSTAIN, C. (1989) *Fuelwood stumpage: considerations for developing country energy planning; industry and energy dept*. Working Paper-Energy Series Paper no. 16. The World Bank, Washington DC.

MITCHELL, C.P., ZSUFFA, L., ANDERSON, S. AND STEVENS, D.J. (eds) (1990) Forestry, forest biomass and biomass conversion (The IEA Bioenergy Agreement (1986–1989) Summary Reports). Reprinted in *Biomass* 22(1–4). Elsevier Applied Science, London.

PHILIP, M.S. (1983) *Measuring trees and forests: a textbook for students in Africa, vision of forestry*. University of Dar es Salaam, Tanzania.

PRIOR, S. (1998) *Personal communication*. Oxford Forestry Institute, Oxford.

RAMANA, V.P. AND BOSE, R.K. (1997) A framework for assessment of biomass energy resources and consumption in the rural areas of Asia. In *Proceed. Biomass energy: key lessons and priorities needs*. IAE/OECD, Paris, pp. 145–157.

ROGNER, H.H. (2001) Energy resources. In *World energy assessment: energy and the challenge of sustainability; Part II Energy resources and technology options* (Chapter 5). UNDP, pp. 135–171.

ROSENSCHEIN, A., TIETEMA, T. AND HALL, D.O. (1999) Biomass measurement and monitoring of trees and shrubs in semi-arid regions of central Kenya. *Journal of Arid Environment* 41, 97–116.

ROSILLO-CALLE, F. (2004) *Biomass energy (other than wood)*. World Energy Council, London, pp. 267–275.

TIETEMA, T. (1993) Biomass determination of fuelwood trees and bushes of Botswana. *Forest Ecology and Management* 60, 257–269.

YAMAMOTO, H. AND YAMAJI, K. (1997) Analysis of biomass resources with a global energy model. In *Proceed. Biomass energy: key issues and priorities needs*. IEA/OECD, Paris, pp. 295–312.

Appendix 3.1   Projecting supply and demand

As indicated in the main part of chapter, there are five major methods for predicting supply and demand, detailed below.

Constant-trend based projections assume that consumption and demand grow in line with population growth and that there is no increase in supplies. It is a useful way to identify any resource problems and possible actions to bring supply and demand into a sustainable balance Essentially, consumption grows with population at 3% a year and supplies are obtained from the annual wood growth and clear felling of an initially fixed stock of trees. However, as wood resources decline, costs will increase and consumption will be reduced by fuel economics and substitution of other fuels.

Projections with adjusted demand is a useful step to examine reductions in per capita demand and its effects in declining wood resources. The adjustments can then be related to policy targets, such as improved stove programmes or fuel substitution.

Projections with increased supplies. Wood supplies can be increased by a variety of measures, e.g. better management of forests, better use of wastes, planting, use of alternative sources such as agricultural residues, etc. Targets for these additional supply options can easily be set by estimating the gap between projected woodfuel demand and supplies.

Projections including agricultural land. In most developing countries the spread of arable and grazing land, together with commercial logging in some areas, is a major cause of tree loss. Land being cleared by felling and burning (*in situ*) would result in greater pressure on existing forest stocks for fuelwood. Using the wood cleared for fuel will contribute to releasing this pressure.

Projections including farm trees. Trees have multiple uses, e.g. fruit, forage, timber, shelter, fuelwood, etc. These farm trees, which are fully accessible to local consumers, are often a major source of fuel in many rural areas and hence should be included in any projection models.

Further reading

LEACH, G. AND GOWEN, M. (1987) *Household energy handbook: an interim guide and reference manual.* World Bank Technical Paper no. 67. The World Bank, Washington DC, pp. 93–94.

RAMANA, V.P. AND BOSE, R.K. (1997) A framework for assessment of biomass energy resources and consumption in the rural areas of Asia. In *Proceed. Biomass energy: key lessons and priorities needs.* IAE/OECD, Paris, pp. 145–157.

YAMAMOTO, H. AND YAMAJI, K. (1997) Analysis of biomass resources with a global energy model. In *Proceed. Biomass energy: key issues and priorities needs.* IEA/OECD, Paris, pp. 295–312.

Appendix 3.2   Techniques for measurement of tree volume

The most common diameter measurement taken in forestry is the main stem of standing trees. This is important because it is one of the directly measurable dimensions from which tree cross-sectional area and volume can be computed. On standing trees the most common diameter measured is the reference diameter usually measured at 1.3 m above ground level and commonly known as the diameter at breast height (DBH).

There are various instruments for measuring at DBH. The most common include callipers, diameter tape, the Wheeler pentaprism, the Bitterlich relascope (or Spiegel Relaskop). The first two are cheap and widely available and the last three are more specialized to foresters and available from specialist suppliers.

Bark thickness

Whether inside bark (i.b.) or outside bark (o.b.) diameter measurements are taken depends on the purpose for which the measurements are made. The proportion of bark volume compared to the total volume with bark varies from a few per cent to about 20% for the majority of species. Bark thickness on standing trees can be determined with a bark gauge.

Calculation of volumes

Using measured volumes

The volume of each tree tallied in the inventory is calculated by substituting the recorded DBH (o.b.) and stem bole length into the following formula:

$$V = aD^b H^c$$

These volumes are tabulated by species and by diameter class; total and averaged for each stratum. The average volumes by species and diameter class are used in calculation of the stock table for each stratum.

Two-entry volume table

If desired, it is possible to develop a two-entry volume table, i.e. DBH and stem bole length values (probably 0.5 m) into the derived formula. The results are tabulated to produce a 'look-up' table showing volumes by diameter class and height class (Figure 3.4).

Average volume tables

Another method of determining average volumes by diameter class is by calculating for some species the average diameter and average height by diameter class, plotting this data as a graph and fitting a least squares curve. Substituting the height and diameter value from this curve into the calculated formula $V = aD^b H^c$ allows you to obtain a table showing DBH class (cm) and volume (m).

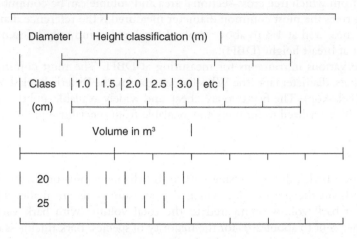

Figure 3.4 Look-up table showing volumes by diameter class and height class.

Further reading

ASHFAQUE, R.M. (2001) General position paper on national energy demand/supply. In *Woodfuel production and marketing in Pakistan*. Sindh, Pakistan, 20–22 October. RWEDP/ FAO Report no. 55. Bangkok, pp. 17–36.

BIALY, J. (1986) *A new approach to domestic fuelwood conservation: guidelines for research*. FAO, Rome.

EMRICH, W. (1985) *Handbook of charcoal making*. D. Reidel Publishing, the Netherlands.

HOLLINGDALE, A.C., KRISNAM, R. AND ROBINSON, A.P. (1991) *Charcoal production: a handbook*. CSC, 91 ENP-27, Technical Paper 268. Natural Resources Council and Commonwealth Science Council (CSC), London

KING, G., MARCOTTE, M. AND TASISSA, G. (1991) *Woody biomass inventory and strategic planning project (draft training manual)*. Poulintheriault Klockner Stadtler Hurter, p. 11ff.

OPENSHAW, K. (1983) Measuring fuelwood and charcoal. In *Wood Fuel Surveys*. FAO, Rome, pp. 173–178.

OPENSHAW, K. (1998) Estimating biomass supply: focus on Africa. In *Proceed. Biomass energy: data analysis and trends*. IEA/OECD, Paris, pp. 241–254.

Appendix 3.3   A short example of charcoal production and use in small islands: a case study of Aassiriki Village, Moso Island and Vanuatu

Ranjila Singh

Tassiriki village locals are engaged in charcoal making. The main purpose of charcoal production in the village is for sale in the market to generate income. Household use of charcoal is rare. Men cut down trees and dry it for burning to charcoal. Old trees and dead trees are also used for charcoal making. It approximately takes 3–4 weeks to dry the freshly cut wood for burning, depending on the tree species and size. When the wood is dry it is burned in big pits. During the burning process, once the charcoal is ready water is poured into the pits to put out the flames. The water is brought from the sea to the charcoal-making site up in the forest by men and women. The pits are covered with iron once the burning is complete so that the charcoal does not get wet from rain. It is left overnight or in some cases for few days to cool down and packed in bags by the women. Men transport the charcoal to the village by in wheelbarrows. At weekends it is transported to the wharf by boats and by trucks from the wharf to Port Vila market. Women are responsible for selling charcoal in the market. They stay overnight at the market house until all the charcoal is sold. Security is present at the market.

People prefer to sell charcoal rather than wood as charcoal has a higher energy content. According to Rosillo-Calle et al. (2007) and Bhattarai (1998) the energy content of 1 tonne of wood is 20 GJ/t (oven dry) and the energy content of 1 tonne of charcoal is 30 GJ/t. The samples of charcoal obtained from Port Villa market showed an average energy content of 27.48 GJ/t. Charcoal is a clean source of fuel as low amounts of smoke are generated while using it and it is less smelly than other biomass fuels (Bhattarai, 1998). It is also easier to transport charcoal than fuel wood as it is lighter in weight and the high energy content thus makes it economically viable (Bhattarai, 1998). The Tassiriki locals use boats to transport charcoal, hence the lower weight of charcoal is easier to tranport than heavy fuelwood; however, some locals still sell fuelwood. Bhattarai (1998) argues that charcoal is a 'high-volume/low-value product' since it requires large space to transport in comparison to its value per unit of weight, a measure in trade indicating that long-distance transportation may not be economical. Other advantages associated with making charcoal are that it has a higher price than fuel wood.

Rosillo-Calle et al. (2007) state that 6–10 tonnes of wood are required to make 1 tonne of charcoal. It was found in the research that 4.3 tonnes of wood was used to make 1 tonne of charcoal, hence the method of charcoal production used here is very efficient.

Further reading

BHATTARAI T.N. 1998. *Charcoal and its socio-economic importance in Asia: prospects for promotion.* Paper Presented at the Regional Training on Charcoal Production. Pontianak, Indonesia.

ROSILLO-CALLE F.P., GROOT DE, P., HEMSTOCK S. L. AND WOODS, J. (eds) 2006. *Biomass assessment handbook: bioenergy for a sustainable environment.* Earthscan, London.

4 Non-woody biomass and secondary fuels

Frank Rosillo-Calle, with contributions from
Peter de Groot and Sarah L. Hemstock

Introduction

This chapter focuses on non-woody biomass, secondary fuels and tertiary fuels. Non-woody biomass includes:

- agricultural crops,
- crop residues,
- herbaceous crops,
- processing residues,
- animal wastes.

And also:

- densified biomass (pellets, woodchips and torrefied wood), which are increasingly being traded internationally,
- secondary fuels [biofuels (biodiesel and bioethanol), biogas, methanol, and hydrogen],
- tertiary fuels such as municipal solid waste (MSW), residues, as their development can have major impacts on biomass resources.

This chapter also assesses animal traction, which still plays a significant role in poor countries and rural areas around the world and impacts on biomass resources.

The methodology employed for estimating non-wood biomass depends on the type of material and the quality of statistical data available or deducible. Where fine detail is required for the implementation of projects, exhaustive investigation is necessary to provide a clear, unambiguous report on each type of biomass resource, and to provide a detailed analysis of the availability, accessibility, ease of collection, convertibility, present use pattern and future trends.

Agricultural and agro-forestry residues as fuel

As stated in the Introduction, there is growing pressure to make better use of existing resources, particularly agro-forestry residues. Most farming systems produce large amounts of residues that offer an enormous potential for energy that is currently greatly under-utilized in many parts of the world. For example, the global residue potential is estimated at about 5 Gt (dry matter) from agriculture and 0.5 Gt from forestry and 6.8 and 0.7 Gt, respectively, by 2030 (Fischer et al., 2007). It is only in wood-scarce areas that raw agricultural residues are often the major cooking fuels for rural households. The greatest concentration of residue burning has been in the densely populated plains of Northern India, China, Pakistan and Bangladesh, where as much as 80–90% of household energy in many villages comes from agricultural residues.

The use of crops residues, however, is changing rapidly. For example, in many areas of China, rapid economic growth and urbanization has led to the rapid replacement of traditional biomass due to fuel switching from crops residues to coal, other fossil fuels and renewables such as wind. This is causing environmental problems such as fire hazards when, for example, residues are left to rot in the fields, or simply burned. In other countries, particularly industrial countries, the energy use of such residues is increasing for modern applications. For example, in the UK almost all straw residue surpluses are now used in combustion plants to generate heat and electricity.

Determination of yield from crop residues

Accurate estimates of the availability of crop residues require good data on crop production by region or district. If this data is not available, a survey will be necessary. A survey should include information on all the uses for crop residues besides fuel (burning *in situ*, mulching, animal feed, house building, etc.) so that the amount available as fuel can be calculated.

Agricultural crops are grown either commercially or for subsistence. It is likely that little or no information will be available for subsistence crops, so it will be necessary to collect data (see below). Total production can then be calculated using existing data on the yields of the various crops (although this data is also often of poor quality).

Crop residues are usually derived from parts of the plant growing above ground. Exceptions include groundnuts, and sometimes part of cotton crop residues. Some communities may also burn roots. When measuring crop residues, it is advisable to record the site at which it is produced (in the field, where it may be left standing or cut, in the house or in the factory). This information is important as the further away from the consumption centre the less likely it is to be used. Table 4.1 gives examples of various types of crop residues at various sites.

An assessment of agricultural residues should include the following steps.

1   *Define what you mean by non-woody biomass.* As already indicated, there is no clear-cut division between woody and non-woody biomass. For example, cassava and cotton stems are woody, but as they are strictly agricultural crops it is easier to treat them as non-woody plants. Classification into woody and non-woody biomass is for convenience only, and should not dictate which

Table 4.1 Various types of crop residues from different sites

Crop	Field (standing)	Field (cut)	House	Factory
Subsistence/cash				
Cereals				
Maize	Stover and leaf cob leaves	Cob	Parchment	–
Deep water paddy	Straw (nara)	Straw (kher)	Kher	Husk
Normal rice paddy	Stubble	Straw	Straw	Husk
Millet, sorghum	Straw	–	Chaff	–
Wheat, etc.	Stubble	Straw	–	–
Cassava	–	Stem	–	Waste
Pulses	Stem	–	–	–
Plantain, banana	–	Stem	Fruit Stem	–
Papyrus	Stem	–	–	–
Heather, etc.	Whole plant[a]	–	–	–
Cash crops				
Coffee (dry process)	(Woody biomass)	Cherries[b]	Cherry, husk	–
Coffee (wet process)	(Woody biomass)	–	Cherry, husk	–
Cotton	–	Roots and stems[c]	–	(Tow)
Coconut, palm nut	(Wood)	Frond	Husk+shell	Husk+shell
Nut trees	(Woody biomass)	–	Shell	Shell
Groundnut	–	Stems	Shell	Shell
Sugar cane	–	–	–	Bagasse
Sisal	–	Old plants	–	Waste
Jute, kenaf, flax	–	Waste	–	Waste
Pineapple	–	Old plants	–	Waste
Indirect use				
Grasses (4)	(Grass)	(Hay)	–	–

Notes: Part in parentheses refer to the woody biomass part of the plant.

[a]In some countries heather-type plants are uprooted from upland areas, dried and burned by householders.

[b]Coffee cherries make a good fertilizer.

[c]Cotton stems and roots have to be uprooted and removed or destroyed within 2 months of harvest because of pathogens and nematode problems.

data is collected. Also, it is important to gather data about agricultural crops and residues that are used for fuel only, not the total non-woody biomass production on any given site. Many plants are unsuitable as fuel, and most have alternative uses.

2   *Obtain data on yields and stocks.* For agricultural crops, dependable information on yields and stocks, quantified accessibility, calorific values, and storage and/or conversion efficiencies must be accurately determined. A study of

the socio-cultural behaviours of the inhabitants of the project area will help to determine use patterns and future trends. Note that for crop residues there is usually no stock, and the yield is the amount generated per annum.

3   *Calculate potential quantities of residues.* It is likely that little or no information will be available for subsistence crops, so it will be necessary to collect data, possibly using remote sensing techniques and even by using drones (see Chapter 8). Total production can then be calculated using existing data on the yields of the various crops (although this data is also often of poor quality). See Appendix 2.1 on residue calculations for further details on how to calculate residues.

The amount of above-ground biomass produced by crops is usually one to three times the weight of the actual crop itself. Estimates have been made of these residues in various countries. Unfortunately, the potential of these residues has not been systematically inventoried. As a result, the quantity of residues has been calculated via estimates of the ratio of by-product to main crop yields for each crop type and the relation between crop and by-product, and by multiplying the crop production of a particular year by the residue ratio; i.e. in the case of wheat 1.3 times as much wheat straw is produced compared to the grain yield, depending on the variety.

Another method for estimating crop residues is to use the Crop Residue Index (CRI). This is defined as the ratio of the dry weights of the residue produced to the total primary crop produced for a particular species or cultivar. The total biomass produced by crop plants is usually one to three times the weight of the actual crop. The CRI is determined in the field for each crop and crop variety, and for each agro-ecological region under consideration. It is very important to state clearly whether the crop is in the processed or unprocessed state. For example, in the case of rice, is the husk included in the crop weight? If the residue has other uses besides energy, a reduction factor should be applied. The quantity of biomass from crop residues that is available for fuel is only a fraction of this total, as not all of it will be accessible.

To obtain accurate estimates of residues production it is thus important to have good estimates of crop production by country, region or district. This may entail undertaking surveys, especially in the subsistence sector, to determine production of both crops and plant residues, and should include all possible uses of residues in addition to fuel. If only general estimates of crops residues are required, crop production figures may be obtained from country statistics or UN bodies such as FAOSTAT. However, such statistics may be based on guesses when dealing with subsistence agricultural production, and hence field surveys may be necessary if accurate information is needed (Ryan and Openshaw, 1991; Openshaw, 1998).

4   *Identifying alternative uses.* There are different types of residues (agriculture, forestry, animal, etc.) with very different characteristics and potential end-use applications. Many of these residues are usually under-utilized and in theory there are considerable opportunities to use them as an energy source.

In practice, however, this potential is often not realised, not only due to availability but also due to other factors such level of socio-economic development, cultural practices, etc.

Despite their potential as an energy source, agricultural residues have to compete with other alternatives, particularly with the need to preserve soil fertility, retain moisture and provide soil nutrients, as well as various other uses of which fodder, fibre and fuel are the most common. Socio-economic changes also mean consumer preferences change and whereas in some areas, e.g. China, these residues are of less value as source of energy, in other countries such as the UK they are used to provide modern services.

5   *Record the site of production and use.* When measuring crop residues it is advisable to record the site at which it is produced (in the field, where it may be left standing or cut, in the house or in the factory). This information is important as the further away from the consumption centre the less likely it is to be used.
6   *Calculate the accessibility of the residue.* The quantity of residues actually available for use is only a fraction of this total, as not all of it will be accessible. Accessibility of crop residues depends mainly on the location and the economic value of the residue. The location determines the collection and transport costs. If this cost is greater than the economic value of the residue it will not be used for fuel.

Processed residues

Processed residues such as bagasse and rice husks are often important sources of biomass fuels and under-utilized. First, identify the relevant industries, and the types of waste they produce. Then collect information on the type, composition (solid or liquid), quantity and dispersal (from intensive or extensive sources) of the waste. Statistical data should be available from the industry concerned, from waste collection firms and the like. For important sources of wastes, this data can be verified in the field at a later date. Of greatest interest are intensive industries producing the larger quantities of waste. These sources should first be investigated.

Bear in mind that the use of biomass is continuously changing and what in the past was of little value can become economically important and vice versa. This is illustrated by the increased use of bagasse and poultry litter for fuel.

•   *Sugar cane bagasse*: it is estimated that the worldwide potential of sugar cane bagasse amounts to about 425 Mt, or approximately 663 million barrels of oil equivalent, most of which was wasted in the past (ISO, 2009). The theoretical potential of bagasse has been estimated to be about 220,000 GWh (Appendix 4.1). Many sugar-producing countries have (or are planning) co-generation programmes (e.g. Brazil, India, Mauritius). Thus, the use of bagasse has been transformed from an undervalued residue that needed to be burned inefficiently just to get rid of it, into a source of considerable economic value (Rosillo-Calle et al., 2012). Moreira (2002) has also identified a possible 10,000 TWh/year from a potential area of 143 Mha of sugar cane around the world (this compares with a planted area of 25.5 Mha in 2005). Appendix 4.1 discusses how Mauritius set up its co-generation programme using sugar cane bagasse.

- *Poultry litter:* another good example is the use of poultry litter in combustion plants in the UK. Poultry litter is the material gathered from broiler houses and contains material such as wood shavings and shredded paper or straw, mixed with droppings. As received, the material has a calorific value of between 9 and 15 GJ/t, with variable moisture content of between 20 and 50% depending on husbandry practices. Poultry litter use is expanding rapidly. This represents a new economic, energy and environmental benefit of a resource, brought about mostly by environmental pressures, that in the past was mostly wasted (Rosillo-Calle, 2006).

When dealing specifically with processed residues it is important to prepare flow diagrams showing:

- the point of origin of the waste,
- the quantity of waste available from each site throughout the year, with historical data if possible,
- the composition of the waste,
- the methods by which the waste is collected,
- the means by which the waste is transported,
- the destination and use (if any) or disposal of the waste.

Such a flow chart will allow you to list the most valuable and available wastes in order of their priority as sources of energy. See Case study 9.3 for more information on flow charts.

Animal wastes

The use of animal residues, particularly dung, is declining except in the case of some larger farms that use it to produce biogas, etc. The potential for energy from dung alone has been estimated at about 20–30 EJ worldwide. However, the variations are so large that figures are often meaningless. These variations can be attributed to a lack of a common methodology, which is the consequence of variations in livestock type, location, feeding conditions, etc. Nonetheless, there are some general rules that can be applied to give overall estimates. However, it is increasingly questionable whether animal manure should be used as an energy source on a large scale, except in specific circumstances, for example:

- manure may have greater potential value for non-energy purposes (i.e. if used as a fertilizer may bring greater benefits to the farmer);
- dung is a poor fuel and people tend to shift to other, better-quality biofuels whenever possible;

- the use of manure may be more acceptable when there are other environmental benefits (i.e. the production of biogas and fertilizer), because there are large surpluses of manure which, if applied in large quantities to the soil, represent a danger for agriculture and the environment, as is the case in Denmark;
- environmental and health hazards are much higher than other biofuels (Rosillo-Calle, 2006). However, in some cases animal residues are increasingly being used mixed with straw and other farm wastes in industrial applications.

Are animal wastes worth assessing?

Measuring supply and consumption will not be useful unless dung is an important source of energy. Enquiries, observations and demand surveys will indicate the importance of dung as a fuel. It takes time, and depends on variables that can be hard to assess. Only proceed if animal wastes are a source (or potential source) of energy. Dung is not an important fuel in all countries, although in others, e.g. the Indian sub-continent and Lesotho, dung still remains an important fuel.

If your initial assessment suggests that animal wastes should be surveyed, you should carry out the following steps.

Determine the number of animals

Determine the number of animals from which the dung is obtained. Reliable estimates of animal numbers are often not available. Be aware of the many pitfalls in calculating animal waste for the reasons indicated above (see Case study 9.3).

Calculate the quantity of dung produced

An assessment of animal wastes will involve:

1   a census of animal numbers by species, and by region or district, and
2   an estimate of their average weight.

More refined data will include information on age and gender. Then use 1 and 2 above to obtain estimates of the average quantities of dung produced, preferably over at least 1 year.

The dung produced by fully grown animals is proportional to the quantity of food they eat. Food intake is roughly related to the size and weight of the animal, which will vary according to the particular country or region. But type and quality of feed are also significant. During the dry season both the quantity and the quality of the feed may decrease, resulting in reduction in the wastes produced. Thus both seasonal and regional feed variation must be taken into account when calculating average or total droppings. Weight and moisture content are the most important data to record at each site, particularly the moisture content at the time of burning and whether the dung is mixed with any other biomass such as straw, as this will affect the energy value.

If the region is prone to droughts, or the climate is erratic, estimates over a longer period should be made. The total animal waste available is the product of the total produced (P), taking into account accessibility factors (A), the collection efficiency (E) and the feed variation factor (F).

In practice, it is not usually necessary to calculate production per 500 kg of live weight. The amount of waste produced can be calculated directly from a census, and from information in the literature. Key data can be verified in a field survey. The weight and moisture content of dung should ideally be measured at each production site. Note also the moisture content when the dung is burned. It is also important to record the amount of other biomass which may have been mixed with the dung since this could affect the energy value.

Calculate accessibility

The accessibility factor for animal dung may vary from zero to one. For housed animals, such as pigs in piggeries, and intensively farmed animals, it can be assumed that waste is 100% collectible and accessible. A census should be carried out for all intensively reared and housed animals. The total droppings for these animals can be calculated, and this figure can be taken as representing the potential dung supply.

For extensively farmed animals (the great majority), estimation of accessibility and ease of collection are more difficult. Methods are similar to those proposed for crop residues. The collection efficiency takes into account those animal droppings which – even though they are accessible – cannot all be collected, which may be a large proportion. The collection efficiency is obtained by a survey to estimate the ratio of the amount collected to the total estimated droppings.

Ascertain other uses for the wastes

Dung has many other uses, such as in house building as a binding agent or to coat walls and floors. Some farmers consider dung so valuable as a fertilizer that they do not use it for any other purpose. However, dung loses its value largely as a fertilizer after it is left in the sun to dry for a few days, although it may still be useful as a soil conditioner. These factors make it difficult to assess both the production of dung and its availability as fuel.

As with industrial wastes, it is useful to develop a flow diagram incorporating the following information:

- the size and distribution of animal rearing enterprises,
- the yields of waste over all seasons, with historical data if possible,
- the moisture content when fresh and when collected,
- the present means of collection, uses (particularly on the land) and disposal,
- use the above to calculate the availability for bioenergy.

Tertiary wastes

These wastes are becoming increasingly attractive, not only because of the large quantities available, but also for environmental considerations. Enormous amounts of MSW are generated annually that were largely ignored in the past. Estimates of the global economic energy potential from MSW vary significantly, e.g. from 6 to 12 EJ. However, in the past few years important changes have taken place that make MSW more attractive. MSW is now even recognized by some authorities as a renewable energy resource, although this is still debatable. It is rather complicated to provide reliable figures on the amount of MSW generated because of the differences among countries, and among rural and urban dwellers within counties (e.g. 314 kg/year per capita in Japan, 252 kg/year in Singapore or 170 kg/year in Brazil). However, there is no doubt that MSW is a sizeable resource that can be converted by incineration, gasification or biodegradation to electricity, heat and gaseous and liquid fuels, and thus should be taken into consideration (see Rosillo-Calle, 2004).

Herbaceous crops

The use of herbaceous crops as an energy source is not new, and in some areas of China it has been normal practice. These traditional practices, however, are gradually being phased out as better alternatives become available. However, some new species are now being used for commercial energy production. It is important that these species are attractive to farmers so that they can be harvested annually using the same machinery as food crops. There are many herbaceous crops being considered as suitable candidates for energy. For example, *Miscanthus* (*Miscanthus × giganteus* and *Miscanthus sacchariflorus*), reed canary grass (*Phalaris arundinacea*) and switchgrass (*Panicum virgatum* L.) have extensively been evaluated as potential energy sources in the USA and some European Union (EU) member countries. Bear in mind that such crops are site-specific and thus much will depend on local conditions.

Miscanthus

Miscanthus has been extensively studied and thus considerable knowledge has been gained that has taken this crop from an experimental concept to the verge of commercial exploitation. As an energy crop, *Miscanthus* differs from short-rotation coppice (SRC) in that it can be harvested annually. A further advantage is that all aspects of its propagation, maintenance and harvest can be done with existing agricultural machinery. Long-term yields at the most productive sites have averaged in excess of 18 t/ha/year, with few agrochemical inputs. Most economic analyses indicate that the most viable market for *Miscanthus* feedstock is electricity generation, and in particular in co-firing with coal power plants.

Reed canary grass

Reed canary grass is a C_3 seed-producing species with similar characteristics to switchgrass, and is well adapted to cool and wet conditions. Yields are comparable to SRC, with energy characteristics similar to straw, and thus it can be used as fuel in modern combustion plants. However, more research needs to be done to determine its long-term commercial viability. The main advantages of reed canary grass are its high potential yields, similar caloric values per unit weight to wood and no requirement for any specialized machinery.

Switchgrass

The potential of switchgrass, a C_4 grass, as an energy source has been extensively researched in the USA, where it is very well adapted to most Northern American conditions. This grass can be directly burned or mixed with other fuel sources such as coal, wood, etc., and thus very suitable for co-firing. Switchgrass grows in a thick cover and can reach about 2.5 m high, and so it is not only a good energy source but also provides excellent cover for wildlife and prevents soil erosion. It is a long-lived perennial plant that can also produce high yields on marginal soil with a low establishment cost. It is cold-tolerant, and requires low inputs and is adaptable to a wide geographical areas. However, much still needs to be learned at a practical site-specific level to be able to consider this grass as a significant possible source for energy.

What is important is that the use of herbaceous crops in the future will be in modern applications and thus, unlike traditional biomass energy applications, commercial measuring techniques (mostly from agriculture) will apply.

Secondary fuels (liquid and gaseous)

As explained in the Introduction, the energy market is changing rapidly and this is having a direct impact on secondary fuels, e.g. the use of shale gas in the USA. Secondary fuels obtained from raw biomass are increasingly being used, particularly in transport, and therefore they will play a greater role in providing energy in the future. In fact, some of them are expected to play a key role. However, it is important to keep in mind that most secondary fuels are used in modern applications and so commercial measurement techniques apply.

Biofuels

The rapid increase in the utilization of biofuels in transport (biodiesel and bio-ethanol), has sparked off a huge debate as to the potential implications on food production and impacts on food prices (see Case study 9.4) and this is the reason why these fuels are explored in some detail in this chapter. Also, according to a joint OECD/FAO (2014) study biofuel production and consumption are expected to increase by 50% by 2023. It is important that biofuel production moves to the next phase; that is, use of feedstocks that do not compete with food crops. A further point to keep in mind is that biodiesel and bioethanol are often presented simply as biofuels rather than placing them in their own categories. For example, the BP Statistical Review (June 2013) estimates that total biofuel production in 2012 was approximately 60.2 million tonnes of oil equivalent (Mtoe; www.bp.com/content/), and the EU gives a figure of 11.3 Mtoe for 2013 and 18.3 Mtoe for 2023 without specifying the exact source of the fuel (www.ethanolproducer.com/articles/). As this book is not concerned with statistical information, the interested reader should consult the abundant literature available (see www.globalrfa.com, www.ebb-eu.org/statistics/, www.eurobser-ver.org/, www.unica.com.br).

Biodiesel

The use of biodiesel has increased significantly in many countries, e.g. Argentina, Brazil, the EU and the USA. But what is important to bear in mind that the rapid increase in palm and vegetable oils has not been caused by biodiesel expansion but rather by demand for non-fuel uses, e.g. 80% of total consumption is for the food industry and about 8% is for other industrial applications, compared to approximately 12% used as fuel. This huge expansion has been driven by dietary changes, particularly in China, India and other developing countries in response to improving living standards. Biodiesel is widely consumed in the EU (13.3 Mtoe in 2013), with approximately 22 Mtoe estimated for 2023 (EC, 2013; www.eurobser-ver.org). Many other countries have ongoing biodiesel programmes, which could have significant implications.

Considerable advances have already been made in biodiesel production and use, including diversification of feedstock, process technology, fuel standards ensuring higher fuel quality, better marketing, diesel engine warranties and support for legal measures.

Biodiesel can be used neat (100% or B100), or in various blends. It can be used in any diesel engine with little or no modification to the engine or fuel system and does not require new refuelling infrastructure. Biodiesel maintains the same payload capacity and range as conventional diesel, and provides similar or slightly reduced horsepower, torque and fuel economy. Biodiesel-like fuels are also being tested as aviation fuel as replacement for jet kerosene, opening up a large potential new market.

Bioethanol

Bioethanol (ethanol, ethyl ethanol; C_2H_5OH) is the most important alternative to road transport in the short to medium term. In 2013 about 85 billion litres (Bl) of ethanol were produced worldwide, which reduced greenhouse gas (GHG) by 100 Mt. The USA, with 56.33 billion l, and Brazil, with 26.6

billion l, are the true giants of bioethanol. The interested reader can consult the abundant literature (e.g. www.globalrfa.com, www.ethanolrfa.org/grfa, www.unica.com.br).

Ethanol can be produced from any sugar-containing material and is currently produced from more than 30 feedstocks. However, in practice only a handful of raw materials are economically viable or close to being so. The principal feedstocks are sugar cane (Brazil) or molasses, and starch crops such as corn as in the USA. In the future one of the most promising feedstocks will be cellulose-containing material if production costs can be significantly reduced.

There are various types of ethanol, as follows:

- biological or synthetic, depending on the feedstock used,
- anhydrous or hydrous, and denatured or non-denatured,
- industrial, fuel or potable, depending upon its use.

There are many sources from which ethanol can be produced, although as indicated only a few are commercially or near-commercially viable. There are two main sources:

- biological, derived from grains, molasses, fruits, etc. (any sugar-containing material),
- synthetically derived from, e.g., crude oil, gas or coal.

Although ethanol can be produced from various different types of raw material, chemically the ethanol produced is identical. Also, be aware that there are two types of ethanol fuel:

1 *Hydrous* (meaning *water-containing*) ethanol (usually 2–5% water) is used in neat ethanol engines (engines adapted to use 100% ethanol). This is possible because, unlike when blended with petrol, there is not phase separation. Only one country, Brazil, has produced large-scale neat ethanol vehicles; however, due to a combination of reasons (e.g. the introduction of the multi-fuel engine) the neat ethanol car is being phased out in favour of flex-fuel (a 'flex-fuel' engine refers primarily as an Otto cycle engine modified to use petrol and ethanol blends in various proportions).
2 *Anhydrous* (meaning *water-free*, or *absolute*) ethanol is blended with petrol in different proportions (flex-fuel). This is the preferred alternative in most countries.

Note also that ethanol can be:

- *denaturized ethanol*, used as a fuel. This is achieved by adding a small percentage of foreign material (can be petrol or other chemicals) that is difficult to remove, to make it undrinkable,

- *non-denatured ethanol* (or 'potable alcohol'), which is the ethanol contained in alcoholic beverages; it is also the starting material used in the preparation of many industrial organic chemicals.

See also Appendix 10.6 on measuring sugar and ethanol yields.

Biogas

Biogas, a mixture of methane and carbon dioxide, is the most important gaseous biomass fuel. Biogas is produced in a digester by anaerobic bacteria acting on a mixture of dung and other vegetable matter mixed with water. The supply of biogas can be estimated by counting the number and capacity of digesters in a district or region, noting how many of them are actually functioning, and any variations in production through the annual cycle. A quick count of digesters can be made if necessary and conversion factors applied to give an estimate of production (see Case study 9.2 and Appendix 10.4).

Biogas production and use can be grouped into three main categories: (1) small domestic production/applications, (2) small cottage industrial applications and (3) industrial production/uses. A significant change in biogas technology, particularly in the case of larger industrial plants, has been a shift away from energy alone toward more 'environmentally sound technology', which allows the combination of waste disposal with energy and fertilizer production, in both developed and developing countries. This has been helped by financial incentives, energy efficiency advances, dissemination of the technology and the training of personnel (Rosillo-Calle, 2006).

While traditional village-like biogas plants (e.g. in China, India and Nepal), are declining, the number of new plants is nevertheless growing as many countries now consider the production of biogas from MSW, landfills, waste, etc., a viable alternative for the reasons stated above. Biogas, which was produced primarily on a non-commercial basis, e.g. for cooking, heating and lighting, is now also produced and used in large commercial plants around the world (e.g. for transport, heating and electricity generation, etc.), although often on an experimental scale. Future biogas applications will be primarily in modern rather than traditional uses.

One advantage of biogas is that it can use existing natural gas distribution systems, and can be used in all energy applications designed for natural gas. But a major disadvantage is its low calorific value. Currently, one of the most common uses of biogas is in internal combustion engines to generate electricity.

Biogas is also compressed for use in light- and heavy-duty vehicles. However, only a few thousand vehicles are thought to be using biogas in comparison to over 1 million using compressed natural gas. There are many experimental programmes around the world. In the past decade important breakthroughs for biogas production and technologies have been made in many countries, particularly in industrial nations. For example, biogas is often used in rubbish/waste collection in an increasing number of countries. However, the main driving force in biogas production is not energy, but the necessity of addressing environmental and sanitary problems. Thus biogas, rather than an alternative energy source, should be considered even more as a potential solution to environmental problems posed by handling of excess manure, water pollution, etc.

Biomethane

Biomethane is an important emerging alternative, particularly in Europe (Germany, the Netherlands, Sweden and Switzerland) where most of the plants are located. Biomethane can be obtained via two main routes: (1) by the upgrading of biogas from anaerobic digestion [manure, wet biomass, industrial waste, etc., and solid biomass (energy crops, corn, grasses, etc.)] and (2) by the treatment, synthesis and upgrading of gas from thermo-chemical treatment of solid biomass (wood and wood residues, SRC, straw, etc.). This option is still in the demonstration stage and is attracting considerable interest (DBFZ, 2012).

Biomethane is a highly flexible energy carrier that can not only be produced from a variety of biomass feedstocks but can also have multiple applications, e.g. as a natural-gas substitute or in combined heat and power (CHP); it offers the opportunity for upgrading biogas plants and, very importantly, it can be used in existing national gas grids. This alternative is becoming quite attractive and should not be overlooked (see www.bioenergytrade.org/docs/biomethane).

Producer gas

Producer gas is generated by pyrolysis and partial oxidation of biomass resources such as wood, charcoal, coal, peat and agricultural residues. It is a practical and proven fuel that has many applications, e.g. transportation fuel, as a boiler fuel for steam and electricity generators, and other industrial uses. This technology is receiving increasing attention as it has proved particularly useful in many rural areas of developing countries, both as a fuel for power generation (particularly co-generation) and in heat production. However, the production and running costs of producer gas continue to be high and applications would make sense only in niche markets, and thus will not be considered in any more detail in this handbook.

Methanol

Methanol (CH_3OH) is currently produced mostly from natural gas and also from coal, but recently there has been considerable interest in its production from biomass. Obtaining methanol from biomass requires pre-treatment of the feedstock and conversion to syngas, which then has to be cleaned before being converted into methanol. This further increases production costs. Using this process, methanol can be obtained from the distillation of hardwoods at high pressure and a temperature of about 250°C, which is in itself a highly energy-intensive process.

Methanol is a common industrial chemical that has been commercialized for over 350 years, and which is widely used, primarily as a building block for thousands of products ranging from plastics to construction materials. Methanol has also been used as an alternative transport fuel blended in various proportions in many countries and is currently under consideration for wider use. Its main appeal is as a potential clean-burning fuel suitable for gas turbines and internal combustion engines, but particularly for new fuel cell technologies (www.methanol.org/).

However, it is the potential role of methanol to provide the hydrocarbon necessary to power fuel cells (fuel cell vehicles; FCVs) that currently makes it attractive. In 2000, the world production capacity of methanol was just under 47.5 billion l. However, since today's economics favour natural gas, methanol from biomass remains a distant possibility unless it can be produced from other sources more competitively.

Methanol also offers few advantages over natural gas, except that it is a liquid and is easier to use in a car. But the loss of energy in the conversion of methane to methanol results in lower overall efficiency and higher overall CO_2 emissions than natural gas when used directly as fuel. In addition, the high toxicity of methanol makes it less attractive as a motor fuel. Thus, on a worldwide basis the methanol market will remain relatively small, mainly for specialized applications such as chemicals and fuel cells, unless oil prices rise considerably.

Hydrogen

Hydrogen has been used for transport since the late 17th century (hydrogen was used by the Montgolfier brothers to power their balloons). However, it was not until 1920s and 1930s that hydrogen was considered truly as a transport fuel. In fact, vehicles using pure hydrogen have been designed and built for over 50 years (e.g. see http://pureenergycentre.com/hydrogen-cars/).

In recent years hydrogen has been researched in many countries as a potential fuel for transport. Hydrogen is believed by some experts to be a major fuel source for transport in the future, with its main potential in vehicles powered by fuel cells, although it is also a perfect fuel for the conventional petrol engine. Hydrogen is not an energy source but an energy carrier and thus requires sources of energy in exactly the same way as the other major energy carriers such as electricity. Like electricity, the advantage of using hydrogen as fuel, as far as security of supply or GHG emissions are concerned, depends on how the hydrogen is produced.

Despite the large potential of hydrogen as a motor fuel, it seems obvious that the advantages of hydrogen will only be achieved after further successful technological developments in storage, transportation, fuel cell technology, and production and distribution facilities. This will require costly investment that would have to be balanced out with other equally potential alternatives, e.g. biofuels, natural gas, etc.

Densified biomass: wood pellets, chips and torrefaction

Densification of biomass is a means of changing waste or low-value biomass products into products that can be used as fuel principally by industry or households, although there are other non-energy applications, e.g. flooring, furniture, etc. There is no question that densified biomass is acquiring increasing importance because of the growing domestic industrial applications for heating, CHP and electricity generation in many countries.

There are various forms of densified biomass but currently the favoured form are pellets which have received considerable attention in recent years, in particular by utilities, in detriment of woodchips (see Appendix 4.2), but there are also many small-scale applications. The most important material used includes agro-forestry residues such as sawdust, wood shavings, rice husks, charcoal powder or fines, sugar cane bagasse and also tall grasses. This process not only facilitates their use as energy but also reduces transport costs significantly.

Wood pellets

In 2010 the estimated global wood pellet production was 14.3 Mt and consumption was 13.5 Mt, with the EU representing 61 and 85% respectively and an estimated installed capacity of about 28 Mt (Goh et al., 2013). Pellets can be classified according to the amount of ash they produce when burned: (1) premium, when ash content is less than 1% (used mostly in small domestic applications) and (2) standard, when the ash content is between 1 and 2% (used in medium-scale applications).

There are four main markets for wood pellets: (1) the small-scale residential market (household heating boilers, up to approximately 100 kW), (2) the middle-scale market (buildings, small-scale industry, community heating and heating plants, with boilers with capacities ranging from about 100 kW to 5 MW, (3) the large-scale market, e.g. large plants and industries, with boilers with capacities of 5 MW or over, and (4) the co-firing market (used primarily in co-combustion in coal power plants) where large amount of pellets can be used, depending of the characteristics of each plant.

The key market drivers can be summarized as follows: co-firing with coal in power plants (potentially the largest market), the development of heating systems (all types), potential availability of sustainable feedstocks in large quantities, development of a global market, the gradual improvement of logistics and support policies which, in most cases, play a pivotal role in market development; policies, however, differ significantly depending on the country or region.

Woodchips

In recent years the market for woodchips have declined significantly and only about 10% all traded woodchips are used for energy purposes. This is for the following reasons:

- higher quality of pellets over woodchips,
- higher overall and transport costs of woodchips,
- health impacts, e.g. the need to take phytosanitary measures (against pine beetle and nematodes) which increases woodchip end prices and limits their use to the higher-priced segments, primarily pulp and paper production,
- countries with large pulp and paper industries are converting their residues/waste into pellets,
- the current availability of shipping capacities for large international intercontinental trade indicates that this is also an important limiting factor for the significant international woodchip trade.

There have been important advances in pellet production which have resulted in significant cost reduction (see Appendix 4.2 for further details). There is ample literature to which the interested reader has easy access (e.g. www.pelletsatlas. info, www.bioenergytrade.org/publications, www.pelletcentre.info/).

Torrefaction/torrefied pellets

Torrefaction technology is not new, but is currently the object of considerable attention by the scientific and trading community. Although currently there are no large-scale plants in operation, it is quite probable that in the very near future (2–3 years) torrefaction technology will be commercially available. Such developments could have important implications for bioenergy (e.g. costs could be similar to those of wood pellets), for the GHG balance, which could be more favourable (especially for long-distance shipping; torrefied pellets have lower GHG emissions due to their higher energy density), and for feedstock availability (e.g. due to the pre-treatment step, biomass feedstocks may become commercially feasible as transport costs could be reduced significantly alongside handling, and storage). Such developments may shift the balance towards torrefied material.

The use of torrefied biomass increases substantially the potential share of co-firing in standard coal power plants (up to 100% in comparison to about 10% based on wood pellets) and allows co-feeding of woody biomass in industrial-size coal gasifiers. Theoretically the torrefaction process can be applied to all kinds of biomass since its main constituents, of hemicellulose, cellulose and lignin, all react during the torrefaction process. Today, torrefaction targets several sources of feedstock: woody biomass from mills, urban waste, forests and plantations, as well as agricultural biomass, residues and energy crops. There are a number of initiatives at different development stages currently working on torrefaction technologies, mainly based in Europe or North America. Torrefaction plants are likely to be built close to biomass resources, e.g. in tropical countries such as Brazil and Indonesia, and most of sub-Saharan Africa.

Animal draught power

Work performed by animals and humans is the fundamental source of power in many developing countries for agriculture and small-scale industries. Animals and humans provide pack and draught power, carry head loads and transport materials by bicycle, boat and draught. In the industrial countries, the work formerly done by human and animal draught power is now mostly carried out by all kinds of machinery. Further details are given in Appendix 4.3.

Draught animals are an important source of power for millions of people. There are about 400 million draught animals in developing countries that provide about 150 million horse power (HP) annually. To replace this with petroleum-based fuels would cost hundreds of billions of dollars. Animal power varies considerably; for example, in most African countries about 80–90% of the population depends on manual and draught power, while the benefits of mechanization serve barely 10–20%. In India between 50 and 60% of the energy used in the agricultural sector is provided by draught animal power.

Thus, although animal draught power is gradually being replaced by machinery, it still constitutes a major source of motive power in the rural areas. Draught animals still offer advantages over mechanical draught power. For example, animals:

- have less impact on the land,
- can work in hilly and inaccessible areas, are 'fuelled' with feed and which can be produced locally, thus increasing fuel independence of the farmer,
- produce manure as a waste product, which is a valuable fertilizer,
- are cheaper to purchase and maintain than heavy machinery,
- are the sources of many other products (i.e. milk, meat, etc.).

One of the disadvantages of draught animals is that they need land on which to be kept and to grow their food. This is a problem for many very small farmers. All kind of animals have been used for draught power, such as horses, bulls and even dogs (see http://en.wikipedia.org/wiki/Animal_traction).

Access to draught animals gives poor farmers in developing countries the means to accomplish the most power demanding farm tasks: primary land preparation such as ploughing, harrowing and transport, and many industrial operations such as crushing and grinding. The alternatives to draught power are often tractors or other equipment, which in many developing countries are too expensive for small farmers, artisans, etc. Another alternative may be manual labour, which may restrict farm productivity and may be unpleasant and arduous.

Draught animal power should not be seen as a backward technology in conflict with mechanization and modernization. On the contrary, given the widespread use of draught power and the fact that it will be the only feasible and appropriate form of energy for large groups of people for the foreseeable future in many countries, it should be given prime attention.

The working performance of a draught animal is primarily a function of its weight, provided it consists of muscle rather than fat. For example, oxen can be expected to exert a draught force of about 10% of their body weight, and horses about 15%. In terms of efficiency, the majority of draught animals produce 0.4–0.8 HP on a sustained basis. However, poor dietary levels often result in draught animals delivering far less power than they would be capable of with proper feeding. An increase in the productivity of the estimated 400 million draught animals in use may possibly raise the productivity of small farms. It is important that farmers recognize this potential for improvement in performance and actively seek increased draught animal productivity.

Performance in field conditions of both mechanical and physiological parameters can be measured with varying degrees of sophistication. The draught force can be measured with a spring balance, hydraulic load cell or electronic strain gauge cells. The distance travelled is simply measured with a tape measure, and speed with a stop watch. Farmers are very adept at judging the degree of fatigue of their draught animals. Nevertheless, some researchers have suggested a more sophisticated system based on respiration rate, heart rate, rectal temperature, leg coordination, excitement, etc. (see Appendix 4.3 for further details).

Future options

All forms of residue are bound to increase in importance for various reasons:

- they are currently an under-used and undervalued resource,
- they are becoming economically and environmentally more attractive,
- they are, in most cases, a readily available alternative,
- there is growing pressure to use existing resources more efficiently,
- supporting policies together with the development of international bioenergy trade should also assist in the commercialization of residues and wastes.

Greater understanding of ecological issues, current and potential energy potential and the economics of using agricultural residues is long overdue, and thus methodological assessment improvements will occur.

Animal residues (particularly manure) are likely to play a diminishing role in a modern world, except when used in modern applications. It is likely that the large-scale use of these residues as a source of energy will be justified only in specific circumstances, for example in locations where a combination of environmental, health and energy use is important. Modern farms use different methods to utilize such resources, e.g. biogas/biomethane production. The methods for assessing manure suffer from many shortcomings due to large variations in animal size, feeding methods, etc.

The most promising are tertiary residues (e.g. MSW), which are increasingly becoming an attractive option around the world, for environmental and social reasons, and for which more refined assessment methods will be required.

Herbaceous energy crops are bound to play a significant role partly because they can be harvested annually and are very attractive to farmers, who can basically use the same machinery as for food crops. Such crops can also be grown on land unsuitable for food crops. In this case, modern agricultural methods will apply.

Secondary fuels (ethanol, biodiesel, etc.) are already used on a large scale in some countries. However, much needs to be learned on the potential effects of the methods of their large-scale production and use. Large-scale development of such fuels will have major impacts on biomass resources and there is a need to address the growing concern with regard to land use competition (see Chapter 7 and Case study 9.4).

Densified biomass, as with secondary fuels, is already being used as a major source of energy, primarily in some countries of Europe. Still missing are internationally agreed standard measurement methods, although significant improvements have been made in recent years and are still being made.

Finally, animal power, so vital in the past, is becoming less and less important as many such activities are being replaced by machinery. However, draught animal power offers many advantages and should not be overlooked, but this is a complex issue beyond the scope of this handbook.

References and further reading

ANON (1988) *Wood densification*. Publication no. 838. Extension Service, West Virginia University.

DBFZ (GERMAN BIOMASS RESEARCH CENTRE) (2012) *Biomethane: key research topics at DBFZ*. www.dbfz.de.

EC (EUROPEAN COMMISSION) (2013) *Prospects for agricultural markets and income in the EU 2013-2023*. Agriculture and Rural Development Report, December. European Commission, Brussels.

FISCHER, G., HIZSNYIK, E., PRIELER, S. AND VAN VELTHUIZEN, H. (2007) *Assessment of biomass potentials for biofuel feedstock production in Europe: methodology and results*. IASA, Vienna (www.refuel.eu).

GOH, C.S., JUNGINGER, H.G., COCCH, M. ET AL. (2013) Wood pellets market and trade: a global perspective. *BIOFPR* (7), 24–42.

HIRSMARK, J. (2002) *Densified biomass fuels in Sweden: country report for the EU/INDEBIF Project*. Sverges Lantbruks Universitet.

HORTA, L.A. AND SILVA LORA, E.E. (2002) *Wood energy: principles and applications* (unpublished report). Federal University of Itajuba, Minas Gerais, Brazil.

ISO (2009) *Cogeneration: opportunities in the world sugar industries*. MECAS(09)05, London, April 2009.

KRISTOFERSON, L.A. AND BOKALDERS, V. (1991) *Renewable energy technologies*. Intermediate Technology Publications, London.

LARSON, E.D. AND KARTHA, S. (2000) Expanding roles for modernized biomass energy. *Energy for Sustainable Development* 5(3), 15–25.

MOREIRA, J.R. (2002) *The Brazilian energy initiative: biomass contribution*. Paper presented at the Bio-Trade Workshop, Amsterdam, 9–10 September 2002.

OBERNBERGER, I. AND THEK, G. (2004) Physical characteristics and chemical composition of densified biomass fuels with regard to their combustion behaviour. *Biomass and Bioenergy* 27, 653–669

OBERNBERGER, I. AND THEK, G. (2010) *The pellet handbook.* Earthscan, London (www.pelletheat.org).

OECD/FAO (2014) *Agricultural outlook 2014.* http://dx.doi.org/10.1787/agr_outlook-2014-en.

OPENSHAW, K. (1998) Estimating biomass supply: focus on Africa. In *Proceed. Biomass energy: data analysis and trends.* IEA/OECD, Paris, pp. 241–254.

ROSILLO-CALLE, F. (2004) *Biomass energy (other than wood).* World Energy Council, London, pp. 267–275.

ROSILLO-CALLE, F. (2006) Biomass energy. In *Landolf-Bornstein handbook, vol. 3: energy technologies.* Springer, Germany, pp. 334–413.

ROSILLO-CALLE, F., GUTIERREZ TRASHORRAS, A.J. AND WALTER, A. (2012) International experiences: sugarcane and bioenergy. In *Bioenergy for sustainable development and economic competitiveness: the role of sugarcane in Africa,* F.X. Johnson and V. Seebaluck (eds). Earthscan/Routledge, London, pp. 231–251.

RYAN, P. AND OPENSHAW, K. (1991) *Assessment of biomass energy resources: a discussion on its need and methodology.* Industry and Energy Dept, working paper. The World Bank, Washington DC.

SIMS. B.G., O'NEIL, D.H. AND HOWELL, P.J. (1990) Improvement of draft animal productivity in developing countries. In *Energy and the environment into the 1990s,* vol. 3, A.A.M. Sayigh (ed.). Pergamon Press, London, pp. 1958–1964.

STARKEY, P. AND KAUMBUTHO, P. (eds) (1999) *Meeting the challenges of animal traction. A resource book of animal traction network for Eastern and Southern Africa (ATNESA).* Harare, Zimbabwe. Intermediate Technology Publications. London.

Websites

http://en.wikipedia.org/wiki/Animal_traction
http://pureenergycentre.com/hydrogen-cars/
www.bioenergytrade.org/publications/
www.esv.or.at/
www.methanol.org/
www.pelletsatlas.info/
www.pelletheat.org
www.ruralheritage.com/horse_paddock/horsepower.htm
www.worldwideflood.com/ark/technology/animal_power.htm

Appendix 4.1 Commercial electricity production from sugar cane bagasse: potential and the Mauritius experience

Vikram Seebaluck

The potential of sugar cane bagasse as an energy resource for power generation

Sugar cane, a commercial crop grown for sugar production in more than 100 countries worldwide, is nowadays also considered as one of the most promising agricultural sources of biomass energy. This is mainly attributed to the use of its 'by-products' for large-scale electricity and bioethanol production which makes the crop a versatile resource that offers multiple end products. The by-product of key interest for electricity generation is sugar cane bagasse, the fibrous residue left after juice extraction from cane in the sugar factory. Bagasse is produced voluminously at a rate of around 0.3 tonne for each tonne of cane processed and provides a significant amount of onsite biomass for power generation. It has historically been burned rather inefficiently in sugar factory in-house boilers to supply internal steam and electricity requirements of the sugar-making process, but with opportunities for electricity export to the power grid and the availability of fully deployed technologies for its efficient conversion into electricity, bagasse is increasingly being processed in annexed modern commercial co-generation plants for exporting large quanta of surplus electricity generated after meeting processing captive requirements. Electricity export can conservatively reach 110–150 kWh per tonne of cane depending upon the processing configuration and products manufactured, namely sugar, ethanol, electricity or a combination thereof (Seebaluck, 2009). Such electricity productivity enables the construction of bagasse co-generation plants as independent entities involved in the business of electricity export from sugar cane complexes that are generally comprised of a triplex of sugar-, electricity- and ethanol-production plants.

The global production of sugar cane has gradually increased over the past decade; it rose from 1,379 Mt in 2003 to around 1,833 Mt in 2012 (FAO, 2014). Actual bagasse production is estimated at around 550 Mt on an annual basis which could be potentially tapped for surplus electricity generation. Competitiveness in the global sugar industry is also resulting into larger-capacity sugar factories processing around 1 Mt of cane annually, which generates an adequate amount of bagasse at factory sites for the viable construction of co-generation plants. Compared to other intermittent renewable energy sources, bagasse co-generation plants have the ability to produce a continuous base load and can efficiently and cost-effectively be integrated into the grid. Such projects not only boost the renewable energy share and contribute to energy access or security but equally support climate-mitigation options as well as sustainable development goals. The significant availability of the resource, its generation at point sources, fully matured efficient conversion technologies and competitiveness with other energy resources are key factors among many others that enhance the development of bagasse co-generation projects.

The largest cane producers, namely Brazil and India, have already embarked on a number of commercial-scale bagasse co-generation plants while many other cane-producing countries such as Mauritius, Australia, Guatemala, Kenya, Uganda, Thailand, the Philippines and Reunion are already producing bagasse electricity. The possible or planned expansion in co-generation capacity in these countries and new power plants in others such as Sierra Leone, Columbia, South Africa and Tanzania provides a good prospect for the penetration of this renewable resource. Bagasse availability and co-generation potential have been estimated at alternative scales: 135,029 GWh/year globally (WADE, 2004), a potential of 3,885 GWh/year in eastern and southern Africa (UNEP, 2007), 5,700 GWh/year in Southern African Development Community (SADC) countries (Seebaluck et al., 2008) and 424 Mt bagasse available globally for conversion at alternative-processing configurations (ISO, 2009). It is estimated that the current average theoretical potential is around 220,000 GWh based on modern high-pressure conversion systems, which could be potentially tapped depending on enabling conditions that should be set up to promote such projects.

Sugar cane bagasse properties and technology options for power generation

The energy content or net calorific value of 'mill-wet bagasse' containing around 45–52% moisture as obtained from sugar factories is around 7–8 MJ/kg (Hugot, 1986; Paturau, 1989; Rao, 1997; Rein, 2007; Seebaluck, 2009; Johnson and Seebaluck, 2012), which is roughly equivalent to one-third the energy content of sub-bituminous coal. Due to the inherent low bulk density of the material, it is common practice to continuously burn bagasse obtained from the sugar factory to reduce handling and storage, which is costly in terms of equipment and facilities. However, bagasse can be densified by briquetting or pelletization to produce almost-dry pellets with improved combustion properties and that can be easily stored and transported over long distances (Seebaluck and Thielamay, 2010).

Bagasse co-generation plants use the Rankine topping cycle for power generation, which is conventional and fully optimized. High-pressure and -temperature boilers working at up to 82–87 bars and 500–525°C with condensing extracting steam turbines (or CESTs) are favoured in modern plants to increase the electricity productivity per tonne of bagasse processed. However, lower-pressure and -temperature boilers with extraction cum back-pressure turbines are also used for electricity co-generation (Figure 4.1). For year-round production, multi-fuel systems using complementary fuels such as coal or other biomass such as trash [sugar cane agricultural residues (SARs)] are generally installed.

The amount of surplus electricity generated is highly dependent on the fibre content of cane or bagasse, the technology used and efficiency of the co-generation plant, and the quantity of captive energy (steam, vapour and electricity) supplied to the processing side (sugar factory and/or distillery including the co-generation plant). Based on actual practices of bagasse co-generation in Mauritius, India and Brazil, a set of benchmarks for the electricity productivity at alternative bagasse-processing configurations and captive consumptions has been developed (Table 4.2). A number of options can be adopted to boost these electricity productivity benchmarks, namely by using much higher-pressure boilers (110 bars), improving the physico-chemical properties of bagasse (reduced moisture content below 50%) and co-firing with SARs which have almost the same net calorific value

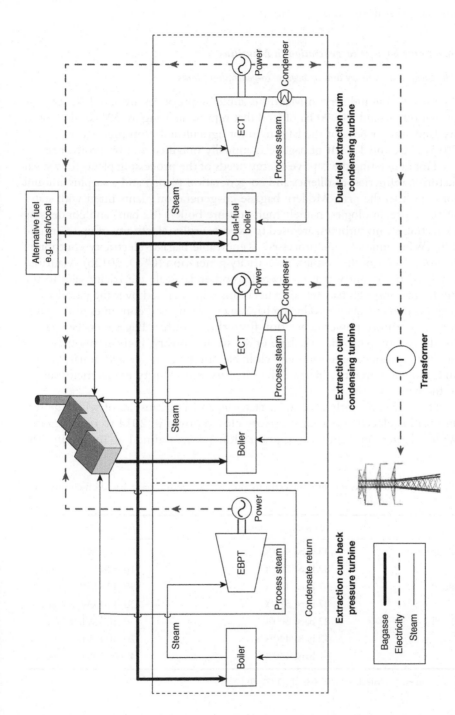

Figure 4.1 Bagasse co-generation options and technologies. Source: Johnson and Seebaluck (2012).

as bagasse. Emerging developments like the use of the bagasse integrated gasification-combined cycle (BIG-CC) or high-fibre cane would further improve the attractiveness of bagasse co-generation plants.

Sugar cane bagasse co-generation in Mauritius

The sugar cane industry and its bagasse co-generation plants

The sugar cane industry in Mauritius annually produces around 4 Mt of cane grown on around 54,000 ha of land that represents roughly 30% of the country land area or 85% of the land used for agricultural crop production (CSO, 2013a). Around 1.2 Mt of bagasse is annually obtained as a by-product that is used for generating the captive energy needs of the processing plants (raw sugar factories, refineries, distilleries and co-generation plants) and a surplus amount for export to the grid. Modern bagasse co-generation plants fitted with state-of-the-art technologies, namely high pressure boilers (82 bars) and condensing extracting steam turbines, are used to export surplus electricity of up to around 130 kWh/tonne of cane processed. Renewable bagasse electricity contributes to around 17% of the country's electricity generation (CSO, 2013b). Mauritius is indeed the country that pioneered commercial-scale electricity sales to the public grid through bagasse co-generation; it has realised over the past decades major achievements in producing bagasse electricity and continues to modernize and optimize power generation from this resource. Bagasse co-generation has the ability to broaden the base of the sugar industry, increase resource productivity, produce renewable and clean electricity, generate additional revenue and, most importantly, increase the diversity and security of electricity supply to the national grid.

There are currently four sugar factories in Mauritius, all of which generate surplus electricity; a sugar factory closed down in 2014 to be integrated with a larger one under a centralization process aimed at improving the

Table 4.2 Benchmarks for surplus electricity production from bagasse based on commercial experience

Country	Operating configuration	Surplus exportable electricity
Mauritius	45 bars, 440°C	60–90 kWh/t cane
	82 bars, 525°C	110–130 kWh/t cane
India[a]	67 bars, 495°C	90–120 kWh/t cane
	87 bars, 515°C	130–140 kWh/t cane
Brazil[a,b]	22 bars, 300°C	0–10 kWh/t cane
	42 bars, 440°C	20–30 kWh/t cane
	67 bars, 480°C	40–60 kWh/t cane

Data sources: [a]Seebaluck et al. (2008); [b]ISO (2009).

competitiveness of the sugar factories. Three of the sugar factories are each annexed to a firm independent power plant (IPP) which operates throughout the year with bagasse during the crop season and coal as a complementary fuel during off-crop season. The remaining one sugar factory of relatively smaller capacity (around 350,000 tonnes cane/year) is a continuous power producer that generates electricity from bagasse only during the cane-crushing season. The total installed capacity for bagasse power generation is 240 MW, compared to the country's installed capacity of around 768 MW (CSO, 2013a). Two of the firm IPPs, of capacity 90 and 70 MW respectively, operate at 82 bars and 525°C while the remaining one of smaller capacity (37 MW) operates at 44 bars and 440°C. Around 350 GWh of bagasse electricity is generated annually while the maximum potential is estimated to be around 450 GWh based on the actual amount of bagasse produced in the country. The additional potential would mainly arise by upgrading the co-generation plant (operating a 44 bar boiler) with a new boiler working to at least 82 bars.

Policies and strategies for bagasse co-generation

The development and implementation of appropriate policies and strategies have been instrumental in tapping the bagasse electricity potential which has been addressed in a number of sequential plans and reforms over the past decades. It started in the Sugar Sector Action Plan (1985), wherein it was proposed to improve energy use in factory processes to save more bagasse for storage and for year-round use. The creation of independent power companies (bagasse-fired power plants) not necessarily linked to the sugar factories was first recommended in this plan. Fiscal incentives were then proposed in the Sugar Industry Efficiency Act (1988) in the form of performance-linked rebates and exemption of income tax for bagasse saved and sold by the larger number of smaller factories (19 at that time) to power plants to ensure the adequate availability of the resource for power generation. The power producers were exempted from income tax to a certain extent for their electricity sales and were given allowances for energy-efficient machinery and equipment used for safeguarding the environment.

The Bagasse Energy Development Programme (BEDP) (MANR, 1991) was the overriding strategy for promoting bagasse co-generation. It was aimed at optimizing bagasse use for displacing the heavy investments on fossil fuel plants to be then effected by the national utility company and reduce the country's dependence on petroleum products. It also aimed at modernizing and improving the viability of the sugar industry, to diversify the energy sector and save the foreign exchange used for petroleum product importation, and contribute in mitigating GHGs by displacing the fossil fuels. Firm IPPs annexed to sugar factories as well as continuous power producers were recommended and the concept of satellite factories supplying excess bagasse to the power producers was adopted. Coal was proposed to be used as complementary fuel during off-crop season because it was cheaply and largely available from supplying countries less exposed to political risks and instability. The necessary man-power was developed and a Sugar Energy Development Loan was negotiated to facilitate implementation of the plan. However, erection of the firm power plants was slow given the inadequate amount of bagasse generated at the low-crushing-capacity sugar factories to which they would be annexed; they had to rely on the supply of bagasse from satellite factories which looked for higher bagasse prices (to be based on coal prices). The minimum capacity of power plants was also found to be 30 MW for a reasonable return on investment and the kiloWatt hour price indexation had to be agreed. Following the experience acquired, three major issues were addressed to facilitate the implementation of the BEDP, namely the funding and fiscal framework (tax-free debentures on equipment and performance-linked rebate on export duty for electricity sales), the consolidation of cane milling activities through centralization (closure and integration of smaller capacity mills into larger ones to reduce costs of produc-tion and generate an adequate amount of bagasse at point sources) and the indexation of the kiloWatt hour price (a price-setting mechanism based on investment in new or second-hand equipment and indexation to the coal price, cost-of-living indices in the country and foreign exchange rate fluctuations).

Strong emphasis was laid on the BEDP via the Sugar Sector Strategic Plan (SSSP) (MAIFS, 2001), which also aimed at maintaining and improving the viability of the sugar industry, reducing the cost of production and improv-ing the revenue generated from sugar and its by-products. Several initiatives were undertaken and it was proposed to continue centralization by placing high priority on electricity generation during the process. It was also pro-posed to adopt energy-conservation devices to reduce captive requirements with the aim of increasing exportation to the national grid. In the *Roadmap for the Mauritius Sugar Cane Industry for the 21st Century* (MAIF, 2005) it was intended that by 2015 IPPs would be located on all surviving sugar factory sites and they would be expected to export about 600 GWh to the grid. Cane trash and more particularly the development of 'energy cane' and 'fuel cane' would be expected to produce high biomass supply to reduce the use of coal. The Government consolidated this plan in a new one, namely the Multi-Annual Adaptation Strategy (MAAS) 2006–2015 (MAIF, 2006), to restructure and adapt the sugar industry to the erosion of preferential mar-kets with respect to sugar. The centrepiece of this latest action plan is the establishment of four sizeable and efficient sugar cane clusters which would remain operational in the long run. The sugar mills are planned to be con-verted to flexible state-of-the-art installations to produce different types of sugar and to optimize the use of bagasse, molasses and eventually cane juice. Many of the planned objectives have already been achieved.

Potential other uses of sugar cane bagasse

Bagasse essentially consists of a mixture of hard fibre with soft and smooth parenchymatous (pith) tissue of a highly hygroscopic nature and it contains mainly cellulose, hemicellulose, pentosans, lignin, sugar and minerals. These constituents provide useful inputs for the production of a number of other products besides electricity production. It is equally important to highlight that an approximately equal amount of fibres as that present in bagasse (factory product) is also available in SARs which is generally left in fields (field product). The latter has roughly similar physico-chemical and energetic properties as bagasse and is available at a rate of 0.3–0.4 tonne of SAR per tonne of cane harvested (Seebaluck and Seeruttun, 2009).

The commercial value-added products manufactured from bagasse are pulp and paper, market pulp, particle boards and fibreboards, corrugated boards and boxes, and industrial chemicals (furfural). Other products that are still being explored are cellulosic bioethanol, carboxymethylcellulose, dissolving pulp, xylitol, biogas and producer gas, biodegradable plastics, charcoal and activated carbon, among others. SARs are used as cattle feed and mulch while their potential uses include the production of electricity, bioethanol, proteins, char, pulp and paper. The different co-products that can be manufactured from bagasse are given in Figure 4.2.

Apart from electricity generation and the production of a few other products like pulp and paper, the cost-effective transformation of cane fibre resources into other value-added products is yet to be commercialized on a large scale although the technologies are available to produce many of them; key factors such as material availability, scale of production, return on investment, market access, alternative competitive resources and pricing are essential for project development, while research and process development should continue for other new fibre-based products. The production of fibre-based value-added products would maximize revenue and contribute

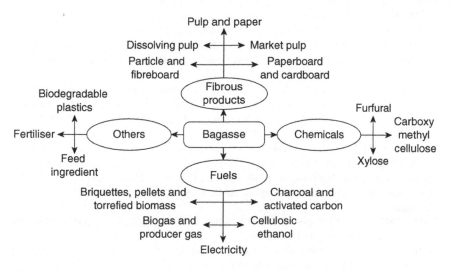

Figure 4.2 Bagasse co-products

to tapping the full potential of sugar cane biomass, thereby consolidating the sustainability of the sugar cane industry.

References

CSO (CENTRAL STATISTICS OFFICE) (2013a) *Digest of agricultural statistics 2012.* Ministry of Finance and Economic Development, Republic of Mauritius.

CSO (2013b) *Digest of energy and water statistics 2012.* Ministry of Finance and Economic Development, Republic of Mauritius.

FAO (FOOD AND AGRICULTURE ORGANIZATION) (2014) *Food and Agricultural Organization statistical database.* http://faostat.fao.org/.

HUGOT, E. (1986) *Handbook of cane sugar engineering,* 3rd edn. Elsevier Science, Amsterdam.

ISO (INTERNATIONAL SUGAR ORGANISATION) (2009) *Cogeneration: opportunities in the world sugar industries.* ISO, London.

JOHNSON, F.X. AND SEEBALUCK, V. (2012) *Bioenergy for sustainable development and economic competitiveness: the role of sugar cane in Africa.* Earthscan from Routledge, Taylor & Francis Group, London.

MAIF (MINISTRY OF AGRO INDUSTRY & FISHERIES) (2005) *A roadmap for the Mauritius sugarcane industry for the 21st century.* Republic of Mauritius.

MAIF (2006) *Multi annual adaptation strategy 2006–2015.* Republic of Mauritius.

MAIFS (MINISTRY OF AGRO INDUSTRY & FOOD SECURITY) (2001) *Sugar sector strategic plan.* Republic of Mauritius.

MANR (MINISTRY OF AGRICULTURE AND NATURAL RESOURCES) (1991) *Report of the high powered committee on bagasse energy development programme.* Republic of Mauritius.

PATURAU, J.M. (1989) *By-products of the cane sugar industry,* 2nd edn. Elsevier Science, Amsterdam.

RAO, P.J.M. (1997) *Industrial utilisation of sugar cane and its co-products.* ISPCK Publishers & Distributors, New Delhi.

REIN, P. (2007) *Cane sugar engineering.* Albert Bartens, Berlin.

SEEBALUCK, V. (2009) *Study of sugar and energy recovery from sugarcane with emphasis on the milling department.* PhD thesis, University of Mauritius, Mauritius.

SEEBALUCK, V. AND SEERUTTUN, D. (2009) Utilisation of sugarcane agricultural residues: electricity production and climate mitigation. *Progress in Industrial Ecology – An International Journal* 6, 168–184.

SEEBALUCK, V. AND THIELAMAY, T. (2010) *Pelletisation of sugarcane bagasse and cane agricultural residues for storage and enhanced energy recovery.* Proceedings of the International Conference on Applied Energy 2010, Singapore, 21–23 April 2010, pp. 1484–1496.

SEEBALUCK, V., MOHEE, R., SOBHANBABU, P.R.K., ROSILLO-CALLE, F., LEAL, M.R.L.V. AND JOHNSON, F.X. (2008) *Bioenergy for sustainable development and global competitiveness: the case of sugarcane in Southern Africa. Thematic report 2: Industry.* European Commission DG-Research FP5 INCO-DEV, ICA-4-2001-10103. Cane Resources Network for Southern Africa (CARENSA)/Stockholm Environment Institute.

UNEP (UNITED NATIONS ENVIRONMENT PROGRAMME) (2007) *Cogen for Africa.* UNEP-GEF Project Document.

WADE (WORLD ALLIANCE FOR DECENTRALISED ENERGY) (2004) *Global report on bagasse cogeneration.* Edinburgh, UK (www.localpower.org).

Appendix 4.2   General aspects of densified biomass and pellets

Frank Rosillo-Calle

As stated in the main chapter, growing interest in densification of biomass has led to a rapid improvement in compacting technologies which have been able to borrow from other industrial sectors, e.g. fodder- and oil-distribution systems. Many types of woody biomass are in principle suitable raw materials for densification. For pellets, for example, to keep costs for drying and grinding low, dry sawdust and wood shavings are predominantly used. The production of pellets from bark, straw and crops is usually more appropriate for larger-scale systems. The requirements for pellet fuel quality vary according to the different characteristics of the wood pellet markets in different European countries.

The following are some additional points regarding raw materials for densification of biomass.

- Densified biomass can vary significantly and has different physical and chemical characteristics. For example, bark pellets have higher ash content than wood pellets and also higher emissions from combustion than wood. Bark, when burned in its raw form, is used in special boilers (e.g. in Sweden where it is very abundant). To mitigate some of these problems, bark may be mixed with wood.
- When tall grasses are used (e.g. elephant grass, switchgrass) they are densified into bales, although they might also be called briquettes. Bales can weigh 250–500 kg. Grasses are becoming increasingly attractive in certain areas which do not have other alternative energy sources and high costs of conventional energy. Grasses are particularly attractive for co-firing with coal.
- Peat is also an organic matter that originates from biomass that does not decompose in the natural environment. Although opinions are divided when it comes to classifying it as biomass or fossil fuel, it is usually considered as non-biomass.
- Densified biomass is obtained in most cases from residues (excluding grasses). The use of energy plantations may be only justified if residues are not available.
- The raw material potentially available for densified biomass is enormous, but costs (energy spent, machinery, etc.) is a major limiting factor.
- Raw material potential for densification of biomass changes with improvement/new technological developments.

Generally, the production of densified biomass fuels entails various steps, as listed here.

- Moisture content is very important and thus the material should be as dry as possible. This means some type of drying equipment may be needed in most cases. Material with a high moisture content is very difficult to condensate.

- Particles should be of a certain size, depending on the end use.
- Conditioning may be necessary, e.g. to make wood fibres softer and flexible by superheated steam to the raw material, to facilitate sizing.
- Decide what to produce as you need different machinery, e.g. flat die pelletizers (for wood) and ring die pelletizers (for bark). For example, in a ring die pelletizer the die is cylindrical and the pellets are pressed from the inside and out through the die by rollers. For briquetting, the dominant technology is the piston press, which pushes the raw material through a narrow press cone by either a mechanically or hydraulically driven piston (see Obernberger and Thek, 2004, 2010).

Costs have been a major obstacle in the past in preventing the use of biomass densification on a large scale, and are key factors to take into account. Costs, together with the high-energy consumption of the process, have improved significantly thanks to new technological advances in this field. These factors have also been assisted by higher oil prices and increasing environmental concerns.

Various studies have looked in considerable detail at the costs of densification, particularly in Austria, USA, Canada, Sweden and the Baltic countries. The costs involved in the densification of biomass can be divided into four main categories:

- costs based on capital and maintenance. These costs are computed using the calculated service lives of the equipment and interest rate. Capital costs are equal to the investment costs multiplied by the capital recovery factor (or CRF). Maintenance costs are percentages of the investment costs, and are calculated on the basis of guideline values;
- usage-based costs, which include all costs in connection with the manufacturing process;
- operational costs, which include all costs in operating the plant, such as personnel costs;
- other costs: this includes insurance, taxes and administration.

Thinking of producing, selling or buying wood pellets?

This may be more complicated than you think because the differing specifications and standard requirements, at national and international level, despite efforts to agree a common methodology. In recent years there have been concerted actions to find common ground on specifications and standardization of woody biomass in general and wood pellets in particular, in an attempt to replace regional and national standards. With so many changes under consideration you are strongly advised to visit the two most important organizations, the International Organization for Standardization (ISO, www.iso.org) and the European Committee for Standardization (www.cen.eu), for the latest information. Here we indicate some of the issues you should be looking for.

Firstly be aware that the ISO alone is currently preparing over 50 standards for solid biofuels (raw material classifications, pellets, etc.) and their various industrial and domestic applications. New ISO EN international standards for fuel specifications are constantly being updated. On the other hand, the European Committee for Standardization has published recently a total of 38 standards for terminology, fuel specification, quality assurance, sampling preparation, etc. The two most important standards under consideration are for classification and specifications (EN 14961) and quality assurance for solid biofuels (EN 15234).

In the specific case of wood pellets, there are some fundamental aspects you need to consider, as shown below.

Specifications

EN 14961-2 specifies a list of testable parameters (normative, which are required, and informative, which are optional). Under this EN specification all specified parameters are normative and include:

- origin and source: requiring the origin of the wood pellets and their composition (according to classification),
- dimensions: minimum and maximum diameter, for all types of wood pellets,
- moisture content: an important parameter that should not exceed 10%,
- ash: also an important parameter, which can vary depending on class, ranging from 0.7 to 3.5%,
- calorific value: another important parameter that must be clearly specified. Net calorific value must be at least 4.6 kWh/kg (class 1) and 4.5 kWh/kg (class 2),
- mineral content: pellets contain a lot of minerals (sulphur, chlorine, copper, nickel, zinc, etc.) which need to be specified. ISO and CEN standards state clearly the maximum allowed for each mineral; to this needs to be added the nitrogen content,
- mechanical durability: pellets are required to have a minimum of 97.5 by weight,
- bulk density: this must be at least 600 kg/m^3 by weight,
- fines and additives: need to be specified too.

To this list you need to add standardization and quality assurance, and technical standards, of which there are many. Chapter 6 also gives some additional details of the schemes available.

Further reading

OBERNBERGER, I. AND THEK, G. (2004) Physical characteristics and chemical composition of densified biomass fuels with regard to their combustion behaviour. *Biomass and Bioenergy* 27, 653–669

OBERNBERGER, I. AND THEK, G. (2010) *The pellet handbook.* Earthscan, London (www.pelletheat.org).

Websites

www.cen.eu
www.duffieldwoodpellets.co/
www.iso.org

Appendix 4.3   Measuring animal draught power

Two instrumentation packages have been developed to monitor draught animal performance under field conditions: an ergometer which measures mechanical and some physiological variables, and a draught animal power logger designed to measure simultaneously both mechanical and physiological parameters so that an animal's responses to various field working conditions can be measured.

There are two main categories of variable: (1) mechanical variables, e.g. vertical and horizontal components of draught force and speed, and (2) physiological variables, e.g. oxygen consumption, heart rate, breathing rate, etc. Attempts to measure these variables have been carried out in both simulated and actual field conditions, e.g. loaded sledges pulled by animals via a force transducer, and comparison of yokes and harnesses.

Comparing power from animals (oxen) with energy from wood

We cannot compare power and energy. We can make a comparison only by specifying a time period (e.g. the time period of an oxen's work). Typically, a good oxen can deliver 0.8 HP, about 600 W. The amount of energy delivered in 1 year by this type of oxen can easily be calculated if one knows the number of hours an oxen works per day and how many days per year. For example, if an oxen works 5 h per day and 280 days per year, the energy from an oxen per year will be:

$$600 \text{ W} \times 1{,}400 \text{ h} \times 3{,}600\text{W} = 3.0 \text{ GJ} \left(\text{about } 1/6 \text{ of a tonne of wood}\right)$$

Comparing human heat and energy

A human body doing little or no physical work needs about 2,000 kcal of energy in its daily diet, which is converted into heat. Thus:

$$2{,}000 \,\text{kcal/day} = 2{,}000 \times 4.2 \text{ kcal/day} = \frac{8.4}{86{,}000\text{s/day}} = 100 \text{ J/s} = 100 \text{ W}$$

Thus a human body doing little or nor work generates a heat equivalent to 100 W. The daily energy in a person's diet is 8.4 MJ. If we assume that the food mainly consists of crops products, i.e. biomass, and considering that dry biomass has an energy content of about 18 MJ/kg, then:

$$\frac{8.4 \text{ MJ/day}}{18 \text{ MJ/kg}} = 0.5 \text{ kg/day of biomass for food}$$

On a yearly basis, the biomass for food per person is:

$$365 \times 0.5 \text{ kg/day} = 180 \text{ kg/year}$$

From surveys, we know that household fuel needed for cooking is about 500 kg/year of dry biomass per person. Hence the ratio of the fuel to food is:

$$500 = \text{approximately } 2.7 \times 180$$

This means that nearly three times as much energy is required under the pot than in the pot!

Further reading

Animal traction, see http://en.wikipedia.org/wiki/Animal_traction, www.ruralheritage.com/horse_paddock/horsepower.htm.

KRISTOFERSON, L.A. AND BOKALDERS, V. (1991) *Renewable energy technologies.* Intermediate Technology Publications, London, pp. 119–132.

LOVETT, T. (2004) Animal power. www.worldwideflood.com/ark/technology/animal_power.htm.

SIMS. B.G., O'NEIL, D.H. AND HOWELL, P.J. (1990) Improvement of draft animal productivity in developing countries. In *Energy and the environment into the 1990s*, vol. 3, A.A.M. Sayigh (ed.). Pergamon Press, London, pp. 1958–1964.

5 The assessment of biomass consumption

Sarah L. Hemstock and Ranjila Singh

Introduction

This chapter examines in detail various methods for obtaining reliable data on biomass energy consumption. It is structured to have particular relevance for the field worker looking at the feasibility of smaller-scale bioenergy projects. The emphasis is on community-level consumption in rural areas of developing countries with respect to the amount and type of biomass resource consumed and that available for project activities. Chapter sections examine suitable assessment methods, appropriate analysis and the assessment of availability of appropriate resources for satisfactory formulation of a bioenergy project. Indicators of changes in biomass consumption over time, which may alter the amount of biomass resource available for future project sustainability, are also examined. Two examples of surveys designed and implemented at community level are given. One example, of a survey designed and implemented in the field (by the non-governmental organization Alofa Tuvalu[1]) to assess biomass consumption in Tuvalu, will be used to illustrate some of the issues discussed in this chapter. The second example is from a research thesis (by Singh, 2013) that focuses on field research in two communities in Vanuatu and three communities in Tuvalu.

Initially, survey design and implementation will be examined since surveys are important tools in determining biomass consumption patterns. Analysis of domestic energy consumption will be investigated because the domestic sector is responsible for a very large proportion of biomass consumption. It provides information on how to measure and analyse biomass consumption. The chapter gives step-by-step details on how fuel-consumption patterns can change over time.

The flow-chart methodology described in the Case study 9.3 is another useful means of calculating bioenergy consumption at various scales: local, national and regional. Analysis of the availability and consumption of the biomass resource is crucial if biomass energy is to be used on a sustainable basis. Calculating biomass energy flows – using reliable data – is one tool which allows consumption to be gauged and predictions to be made concerning availability, consumption and sustainability of biomass energy and also highlights areas in the harvest-to-end-use chain where improvements in efficiency can be made.

Designing a biomass energy consumption survey

Any survey will produce data that is biased, particularly if the data is gathered by questionnaire. Therefore, before embarking on an appraisal it is necessary to establish the objectives of the survey and determine what data is required from it. In particular you should ask the following questions.

- What is the reason for the survey? Consider the advantages and disadvantages of using questionnaires in your particular circumstances.
- What questions need to be answered?

  o Prepare written objectives for the research.
  o Have your objectives reviewed by others.

- Is the survey strictly necessary?

  o Biomass energy use often fails to be included in official government statistics. However, it is worth reviewing any literature related to your objectives since surveys are time-consuming and costly so you do not want to repeat the work of others.

- What actions/interventions/activities are going to be based on the results of the survey?
- What level of detail is required?
- Which sectors will be surveyed?
- What are the resources available? (Example: if you decide on a questionnaire, determine the feasibility of administering it to the population of interest.)
- Once the population of interest is established, gender considerations should be taken into account if it involves both men and women.
- How long is this process going to take? Prepare a time line.

The answers to these questions may suggest a methodology for data collection, compilation and presentation. The essential details can then be expanded where it is necessary, desirable or possible.

Field example 1 Alofa Tuvalu:
Tuvalu Biomass Energy Consumption Survey

The objective of the survey (Hemstock, 2005) was to determine the amount of biomass available for two community bioenergy projects in Tuvalu: pig waste biogas digestion and coconut oil biodiesel. A detailed literature review was undertaken and it was found that biomass energy use was not included in any previous contemporary literature in the detail required. Other domestic energy use [such as liquid petroleum gas (LPG) and kerosene] had been estimated in a recent household survey based on the number and type of cooking appliances. This assessment was not accurate as actual fuel use was not accounted for. Literature detailing standing stock, vegetation type/class, land use issues and house construction was also reviewed. In order to establish the amount of biomass available for these two projects it was deemed necessary to carry out a detailed assessment of energy use in the domestic sector.

The results from this survey could then be considered against annual productivity and standing stock. The best method of carrying out the survey was by questionnaire and by weighing wood piles. The questionnaire survey method was advantageous since households were close together and it was possible for project staff to question each household individually so a fairly accurate assessment of domestic bioenergy use could be obtained. In addition, local organizational infrastructure was engaged and women were questioned at a local women's group and 'Kaupule' (local council) meetings. Language was one disadvantage that had to be overcome using local representatives. A survey team consisting of 10 local representatives (nine of them women) and two project staff (both women) carried out the survey across three of Tuvalu's nine islands over a 6 week period in 2005.

Field example 2 Gender dimensions of household energy and sustainability of woodfuel use

In addition to Hemstock (2005) study this research looked at gender dimensions of household energy access, use and decision making (Singh, 2013). The study also analysed the sustainability of fuelwood use for household cooking. From a literature survey it was found that there are gender dimensions to household energy and in Pacific Island countries there is gap in gender and energy studies. Hence the study was carried in two rural communities in Vanuatu and three communities in Tuvalu (two urban and one rural).

The method used to determine the gender component of household energy was by the use of questionnaire and focus groups. The questionnaire focused on household lighting, cooking and fuelwood collection (see Appendix 5.1). It was ensured that an equal number of surveys was given to men and women. Furthermore, discussion was carried out with men and women separately in the focus groups. The main aim of the questionnaire and focus group was to find out who had access to, who used and who made decisions over household energy parameters such as lighting, cooking and fuelwood collection. The surveys were translated into the local language and a local research assistant accompanied the researcher to the field sites for translation purposes. For the sustainability of fuelwood use survey sheets were used to collect data on household fuelwood use and quadrant surveys were carried out in the fuelwood collection area (see Appendix 5.2).

Questionnaire design

Very often, surprisingly few fairly general questions can capture many of the issues relating to biomass consumption (food and fodder use, construction, domestic energy, fibre for mats and clothing, fertilizer). Surveys/questionnaires are important tools in helping you to form a realistic picture of biomass consumption patterns. You should focus initially on issues you consider absolutely necessary. It is important to underline the preliminary nature of any programme design at this stage. Do not forget that local conditions play an important role in the final shape of the survey.

In order to obtain reliable data you need to:

- group the items by content, and provide a subtitle for each group,
- within each group of items, place items with the same format together,
- indicate what respondents should do next or what the information will ultimately be used for, at the end of the questionnaire,
- prepare an informed consent form, if needed,
- consider giving a token reward for completion of the questionnaire and providing refreshments for focus groups [to encourage participation a gift of vegetable seeds for family gardens was given by Alofa Tuvalu personnel and the survey by Singh (2013) provided refreshments to the focus group participants and fee in form of money was given to the community leaders that was used for community projects such as water tanks],
- consider preparing written instructions for the administration if the questionnaire is to be administered in person (the preferred method), and also prepare focus group guidelines.

Detailed village surveys are a crucial element in the preparation of biomass energy projects. They provide an opportunity for people to express their opinion about the problems they face and how best to solve them, and for local communities to be involved with survey design. It is important that local people are actively involved from the beginning, and that they trust and gain the confidence of those who pose the questions. It is also important to ask the right questions to the right people. For example, if the survey is dealing with household energy consumption then it is wise to ask the woman's point of view as she will be able to provide more accurate information, and any project intervention is likely to affect her more directly.

Outlining required information

Problems can occur when elements of the survey are not strictly defined. It is therefore important that interviewers and interviewees understand exactly what is meant by the survey questions and what analysis is required from the answers. For example, a question as simple as 'household size' is not as easy to define as it first appears, and may not relate to the number of people using domestic bioenergy for their cooking needs. A preliminary (pre-study) survey can be helpful in some cases. This can allow the researcher to gain further understandings on how some questions are seen by the interviewees and the possible response they can get. Accordingly, alterations can be made to the questionnaire for full-scale research.

Example: defining a household

Domestic use often accounts for the largest consumption of biomass fuel in many developing countries. However, it is important to be clear exactly what we mean by 'household' before considering how the sector should be disaggregated. The term household is subject to various definitions, and estimates of per capita consumption will vary according to which one is used. At different times, operative household size may be defined as follows:

- the number of family (or cooking unit) who sleep and eat in the house,
- the number of family who sleep but do not eat in the house,
- the number of family who eat but do not sleep in the house.

Definitions may also include:

- the number of labourers who eat but do not sleep in the house,
- the number of people who work elsewhere, but who regularly eat some meals with the household,
- the number of people naturally included as members of the household (e.g. head of household who is a labourer, but who actually eats elsewhere),
- the number of people for whom food is cooked and taken to, but who actually live elsewhere (e.g. grandparents and children).

There are yet further possibilities. For example, a working member of the family may receive payment in food that is brought home, rather than money. In this case it is important to discover whether the food is cooked or not.

The Alofa Tuvalu survey (Hemstock, 2005) and Singh (2013) accounted for the number, age and sex of people who slept in the house and the average number of people cooked for on a regular basis.

Some broad determinants of energy consumption in the household sector are outlined in Table 5.1.

Field example 1

Question groups used in the Tuvalu survey (Hemstock, 2005) are listed here.

Table 5.1 Determinants of domestic energy consumption

Location	Climate, altitude
Social	Definition of 'household' consumption unit, demographic patterns, income distribution
Economics of supply	Costs in terms of price and collection effort
Cultural	Diet, fuel preferences

- *Household*: size (number of males, females, under 18s sleeping in the dwelling), and the average number of people cooked for each day.
- *Kitchen appliances*: ownership (LPG stove, kerosene stove, charcoal stove, electric stove, wood-burning stove, open fire, electric rice cooker, electric kettle, other), frequency and duration of use for each appliance (boiling drinking water, cooking rice, cooking fish, etc.), number of times each appliance is used per day, per week and per month.
- *Household fuel use*: litres of kerosene used per week, bottles of gas purchased per month/year, connection to grid, amount spent on electricity annually, inclusion in solar energy programme, number of open fires per week, amount and type of wood used per fire (usually coconut husk with some shell), number of times per week charcoal is made (coconut shell charcoal); fuelwood and charcoal was also weighed.
- *Communal cooking*: number of community events per month; an open fire is usually the preferred cooking method for communal events in Tuvalu. Preparation of food for sale (via women's groups providing meals for school children, etc.).
- *Coconut consumption*: number consumed per household per day by humans and animals (pigs and chickens). Coconut residue available for use as bio-energy (husk and shell) can be estimated from this.
- *Manure production*: number of pigs owned by each household, how often are pigs cleaned out, to what use the resulting slurry is put (usually washed into the sea or used to fertilize banana and vegetable plots). This information can be used to estimate the amount of residue for use in biogas digesters.

The survey questions were delivered in person and were phrased so that certain answers could be verified by other answers (e.g. number of times the kerosene stove is used per day and the amount of kerosene purchased per week). This approach gives an indication of the validity of the data obtained. Construction materials used for domestic dwellings were not considered as previous reliable surveys detailed the amount of locally available biomass used for this purpose. In addition, biomass use for brick-making and beer brewing was not considered as this type of activity is not undertaken in Tuvalu. Handicrafts and thatching using plant fibre (pandanas leaf) was also not considered as harvesting for this purpose is not destructive and the materials are used with virtually no biomass residue production.

The results of the survey were used in the formation of a 10 year renewable energy project to implement solar, biogas and coconut oil biodiesel.

Field example 2

The household survey (Singh, 2013) included socio-economic characteristics of households, such as:

- gender (surveys can be analysed on a gender basis to indicate if there is gender dimension to household energy in respective areas),

- main source of household income (can be linked to how much of the income is used to obtain household energy or how much is earned from energy parameters such as sales of charcoal and/or firewood),
- who manages household income [can be related to purchase of household appliances and type of energy used for cooking and lighting whether bought (such as gas, kerosene) or collected (such as firewood) on a gender basis],
- education level,
- household size (number of adults and children),
- major household expenses.

Energy resources: access and management:

- household appliances or goods,
- household lighting:
  - o   main source of lighting,
  - o   cost of lighting per week,
  - o   who manages household lighting;

- energy for cooking:
  - o   main type of energy used for cooking,
  - o   if the energy is bought, what is the cost per week,
  - o   who does most cooking,
  - o   time spent cooking per day,
  - o   who manages the energy sources of cooking,
  - o   if wood or forest materials are used, which species are used,
  - o   if firewood is used, who collects it,
  - o   how much time is used for firewood collection per week,
  - o   is drinking water boiled (if yes, what is the source of energy for boiling),
  - o   if an open fire is used, is it located inside or outside (health implications),
  - o   type of firewood stove used.

When designing a survey to assess rural community bioenergy use in developing countries there are usually four broad questions to consider.

1   *If there is an energy shortage, particularly firewood, how severe is it?* This is an important question since this assessment is the key to any bioenergy project activity or intervention. The energy situation in rural areas may be roughly gauged by a number of ranked indicators (such as localized deforestation). The accuracy of such an analysis is dependent on whether enough detailed, spatially disaggregated information on these indicators is available. For example, one indicator is fuel switching and the types of fuels being used. The absence of a charcoal industry may suggest that fuelwood is still plentiful and able to supply domestic energy needs, since charcoal only

becomes economically attractive when wood becomes scarce and has to be transported over long distances.

2   *Which groups are under the greatest stress?* Once the vulnerable areas and groups of people are identified, they must be visited in the field. This is essential since indicators need to be identified accurately as some results can be misleading and therefore need to be checked. For example, fuelwood use (in terms of mass of fuelwood used) can actually be higher in areas of fuel-wood shortage than it is in areas which are less stressed in terms of indigenous woodland resources. This may be because, firstly, in areas where fuelwood is in relative abundance consumption is lower as fuelwood is collected, stored and well dried before use. Secondly, in areas where fuelwood is scarce, species with poor burning qualities and wet wood may be used without drying, thus increasing the measured weight of fuelwood required (since it is burnt wet). These indicators can only be measured reliably in the field.

In the case of fuelwood and charcoal, factors such as income level and fuel availability are crucial. The switch to dung or crop residues (unless through improved technology) may signify the lack of choice for 'better' fuels and is therefore not voluntary. The use of crop residues and dung for domestic pur-poses may indicate fuelwood scarcity and so is often resented by the majority of its users. Strategies to increase the use of residues must acknowledge these problems and communities must be involved in any project planning process.

3   *What end uses are to be considered?* It is difficult to separate the use of fuel for different household functions such as cooking, heating and boiling water, since the fire is often used for several functions simultaneously, and the fuel store is not differentiated. It is not worth trying to disentangle the amount of fuel used for each activity, particularly in a rapid survey, as this informa-tion is not particularly useful in the design of interventions. However, it may be worth noting in a qualitative way whether there are regular fuel-using activities in addition to cooking, since this might possibly justify the future introduction of a specialized energy-saving device[2]. However, for most purposes, small-scale activities (e.g. water boiling, home-consumption beer brewing, ironing, etc.) are best included under general fuel use.

Other major end uses of fuel at the household level could include cot-tage industries, e.g. commercial beer production and the preparation of food for sale. Outside the household there may be small industrial uses such as bakeries, brick kilns, fish smoking and the like.

Include these activities in the demand survey if some form of interven-tion is possible or if total bioenergy use is required in order to assess the amount of biomass available for other purposes. Try to get details on num-bers of people (women) involved, quantities of fuel used, and the current costs and constraints. If, realistically no such intervention is possible, it is best to simply note the occurrence of the activity and leave it at that.

Systematic methods of physical weighing and measuring are only possible during detailed, long-term surveys, as they require a great deal of time both during and after the survey (e.g. in assessing the moisture content of the sampled wood). This is by the far the most reliable survey method. In the case of Tuvalu the survey took place over 6 weeks; seasonality and climatic variations were not an issue since temperature and rainfall remain fairly constant throughout the year.

Recall by the interviewee concerning the amounts of biomass used is unfortunately not very reliable, because the concept of volumetric or weight measurement of firewood is unfamiliar to most people. The frequency of fuel-gathering trips is a more reliable measure. One simple method is to ask the respondent to make up a bundle of typical size that he/she brings home from a trip, and then determine by questioning how many such trips are made. It is important to differentiate between seasons when estimating the number of trips made.

The bundle should then be weighed. Estimate the moisture content of the wood, as it can account for as much as half of the total weight. You can measure the moisture content of the wood in the field with a portable meter, or samples can be taken back to the laboratory. If it is not possible to make physical measurements, you can estimate the moisture content by stating whether the wood is green, partly dry or dry. This will provide a rough but useful comparison with other households or villages in the same general area.

The above refers to firewood. However, measurement of the use of charcoal, crop residues and dung as fuel has always posed additional difficulties. However, it may be possible to get relative information.

Some of the factors discussed in points 1–3 above may also provide an indication of the minimum needs, as opposed to actual consumption (which may be far above or below the needs) in a specific location or socio-economic or cultural setting; the list is not exhaustive. Most of the key determinants of domestic energy use are interrelated. Moreover, they are frequently perceptions that the questioner will have to elucidate; e.g. it is the perceived 'effort' of gathering fuels within the context of many other tasks, rather than actual distance or time involved that matters.

4   *How big should the sample be?* Identify the accessible population. The size of the sample depends largely on the number of sub-sectors. A very limited number of contacts or sample points are needed within each sub-group. For each group, build up a holistic picture of the prevailing demand through observation, discussion with officials and workers who are familiar with the people, and group discussions with the subjects of the demand survey. It may be possible to focus on the experience of one or two of these groups to draw out the discussion and raise comments. Consider using random cluster sampling when every member of a population belongs to a group. It is important that as many aspects as possible are taken into account when estimating supply and demand of biomass. Avoid using samples of convenience. Simple random sampling can be a desirable method of sampling

under certain circumstances. It is extremely difficult, and often impossible, to evaluate the effects of a bias in sampling.

Consider the importance of getting precise results when determining sample size. In a small community of, say, up to 200 people you should try to get a response from everyone. In order for any survey sample to be statistically significant you must sample at least 5% of the population, but where the population size is small the 5% guideline is not accurate. For statistical analysis, accuracy is determined by sample size alone; however, remember that using a large sample does not compensate for a bias in sampling, so always use random sampling and multi-level analysis techniques (see below). (The bias in the mean is the difference of the population means for respondents and non-respondents multiplied by the population non-response rate.) Make sure that all population groups within the sample area are represented.

Variability within the data generated from the questionnaires should also be assessed. This can be done by using the median as the average for ordinal data and the interquartile range as the measure of variability. Alternatively, use the mean as the average and the standard deviation as the measure of variability. Using the mean as the average is usually the most reliable method for data generated from bioenergy surveys since the standard deviation has a special relationship to the normal distribution that helps in its interpretation. Use the range very sparingly as the measure of variability. For the relationship between a nominal variable and an equal interval variable, examine differences among averages. When groups have unequal numbers of respondents, include percentages in contingency tables. For the relationship between two equal interval variables, compute a correlation coefficient.

Errors can come from:

- small sample size,
- not using random sampling,
- the use of an inadequate time frame,
- a badly designed questionnaire,
- recording and measurement errors,
- non-response problems.

Assessment of sustainability of household fuelwood use

Household fuelwood inventory (see Appendix 5.1):

- select a representative or random sample of households that represents your area of study,
- quantity (for one household meal; pieces of wood, number of coconut shells and husks, etc.),
- species of wood used (major and other minor ones); collect a sample of major species for further analysis in laboratory,
- weight (total).

Quadrant survey (see Appendix 5.2):

- coordinates,
- species name,
- number of same species in the quadrant (tally),
- diameter at breast height (DBH; measure circumference at breast height and later calculate the diameter using the formula: circumference = $\pi \times$ diameter).

The procedure used by Singh (2013) to calculate the sustainability of fuelwood use is discussed below. Three factors are required to calculate sustainability and estimate the number of trees or tonnes that need to be planted annually for sustainability if fuelwood use is currently unsustainable: (1) weight of fuelwood (in tonnes) used per year for each fuelwood species for the whole community or village, (2) standing stock (in tonnes) for each species in the whole fuelwood collection area and (3) mean annual increment (MAI; tonnes) for each fuelwood species in the whole fuelwood collection area.

1   For the household fuelwood inventory a household fuelwood survey was conducted (see Appendix 5.2). Ten households were selected from each of study sites and samples of fuelwood (sample species names were noted) were collected for further analysis. The percentage moisture content of the fuelwood samples was calculated on a weight basis. Energy content of fuelwood samples (three replicates) was calculated using a bomb calorimeter. The procedure was adapted from the Gallenkamp Ballistic Bomb Calorimeter standard operating procedure (https://www4.rgu.ac.uk/files/Gallenkamp%20Ballistic%20Bomb%20Calorimeter.pdf).

The weight in kilograms of fuelwood per day was multiplied by 365 to get the annual fuelwood use per household. The weight (kg/year) was converted to tonnes per year. Different households have different species as their major fuelwood, hence the average tonne of fuelwood use for each fuelwood species was calculated from the sample of ten households. This average value was multiplied by the number of households in the community. Hence the tonne of fuelwood use per year for each species was obtained.

Fuelwood (kg) per day × 365= fuelwood (kg) per year → converted to tonnes per year → average for each species of fuelwood → multiplied by the number of households in the community → tonne of each species used per year

2   An on-ground assessment of standing stock of fuelwood was made for the study sites using standard sampling methodologies. Two quadrat surveys were carried out on the fuelwood collection area. The number of fuelwood trees, their height and circumference were noted. From the tree circumference the DBH was calculated as described above. The diameter was made the subject of the formula: diameter = circumference/π. The standing stock (volume) for each tree of every species was calculated using the volume

formula: (height × DBH²)/12,732 in cubic metres. The average was calculated for each species to obtain the average standing of each species in a
quadrat. Since two quadrats were taken, once all the tree volumes were calculated the average for each species in each quadrat was taken, then the two
quadrat averages for same species were added and divided by two. Hence
the average standing stock (volume in m³) was obtained for each species.

The quadrat area was 400 m² but the total fuelwood collection area was
greater, hence the area for whole fuelwood collection was found. The
total fuelwood collection area was converted to square metres and divided
by 400 m². This gave a factor that was multiplied by the average standing
stock (in m³) of each species in the quadrat in order to get the standing
stock for the whole fuelwood collection area. This standing stock of each
species in cubic metres was then converted to tonnes. Hence the final
standing stock of each species is given in tonnes.

3   After that the percentage MAI (%MAI) of each fuelwood species was calculated. The average tree height of species was calculated. Then %MAI of each
species was calculated using the formula: %MAI = growth rate (m/year)/
average height. After the %MAI of each species was calculated, the average MAI was calculated with the formula: average MAI = %MAI × average
standing stock (m³). Now we have the average MAI of each species this is
for the quadrat only. Hence the MAI for the whole fuelwood collection area
needs to be calculated.

The quadrat area was 400 m² but the total fuelwood collection area was
greater, hence the area for whole fuelwood collection was found. The total
fuelwood collection area was converted to square metres and divided by
400 m². This gave a factor that was multiplied by the average MAI (m³) of
each species in the quadrat in order to get the MAI for the whole fuelwood
collection area. This MAI of each species in cubic metres was then converted to tonnes. Hence the final MAI of each species is given in tonnes.

Calculation of sustainability

Sustainability for each fuelwood species (see also Chapter 6) was calculated with
the formula: sustainability = MAI of the species (tonnes) − fuelwood use of the
species per year (tonnes). If the sustainability figure is a negative value that means
the fuelwood use is unsustainable and vice versa. The unsustainable figure gives
the tonnage of fuelwood trees that needs to be replanted per year to be sustainable. To calculate the number of trees the formula is: Number of trees to be
planted = [sustainable figure for the species (remove the negative sign)/standing
stock for the species] × total number of trees of the species.

The total number of trees should be calculated prior to making these calculations. As the quadrat area was 400 m² but the total fuelwood collection area
was greater, the area for whole fuelwood collection was found, as described
above. The total fuelwood collection area was converted to square metres and

divided by 400 m^2. This gave a factor that was multiplied by the number of trees of each species in the quadrat to get the total number of trees in the whole fuelwood collection area.

Implementing the survey

Before implementing a full-scale survey, consider asking about ten individuals to provide detailed responses to a draft of your questionnaire. This should give you some idea as to the responses you are likely to receive and will highlight any ambiguous areas. If you are planning to 'score' questions on your survey (example: score 5 for strongly agree, 4 agree, 3 no opinion, 2 disagree, 1 strongly disagree), tally the number of respondents who selected each choice, then compare the responses of high and low groups on individual items. If there is a large spread of difference in the responses it usually means that respondents found the question to be ambiguous. Testing your questionnaire in this way will ensure that the larger-scale implementation of the survey will fulfil your research objectives by providing the type and detail of information you require in a format that will be useful.

The following points are important and should be considered for the successful implementation of a survey.

The size and composition of the survey team

A biomass consumption survey is most effectively carried out by a small, multi-disciplinary team, composed of both men and women. In most cases the team in the field should be small (at least two but not more than three people) so as not to overwhelm local facilities and the people being assessed. They should work together, as this facilitates development of a holistic view of biomass consumption, through shared observations. The Tuvalu survey was carried out by two Alofa Tuvalu personnel and ten local representatives.

The number of survey teams

Placing a number of small teams in different areas is usually not very satisfactory, as this inhibits a comparison of the findings. One small team is much better able to develop the necessary sensitivity. The survey by Singh (2013) comprised one local (a research assistant) and the researcher in the Tuvalu and Vanuatu communities.

The length and nature of a survey visit

Initially the team should make short, informal visits to targeted communities to interview and observe. From these 'impressionistic' surveys, the team can decide on the necessity and structure of more detailed surveys. Whatever the type of survey, it is important to obtain the opinions and perceptions of all social groups, particularly women, landless and other disadvantaged groups whose views are often neglected or poorly documented. Team members should maintain a flexible and informal approach by making regular evaluations of the progress of the survey.

The Alofa Tuvalu survey was integrated with local social organizational structures as well as accounting for grass-roots-level bioenergy users via house to house enquiries. It is also important to explain the purpose of the survey; interviewees may then provide more relevant information or reveal aspects of fuel use that the interviewer had not considered. For the Singh (2013) community level survey the communities were aware of the researcher as formal arrangements were made with the community through community leaders.

Methods

Although informality is stressed, general methodologies should be worked out in the office prior to embarking on data collection. The design of the survey should direct the fieldwork so as to maximize the quantity of useful information obtained. You should attempt to understand the ongoing processes that determine consumption and supply. For example, due to Tuvalu's geographical isolation kerosene and LPG supply was sporadic so domestic biomass energy use was higher when other fuel sources were unavailable.

The overall picture

The most effective use of resources would be to combine the consumption survey with the supply survey, as was the case with the Alofa Tuvalu survey. Any calculation of future trends should be undertaken against a background of change.

The compatibility of data

It is important to employ methodologies that allow comparisons between different groups and areas. Careful consideration as to which method to adopt must be given before the survey team reaches the field.

The distinction between assessment and estimation is important

A clear distinction must be made between assessment and estimation. Assessment is an analytical exercise which yields fairly reliable information from the respondent, and that can stand up to the rigours of statistical analysis. This may be done through questionnaire survey. An estimation will deal primarily with physical processes of measurement and quantification of primary data.

Distinguish between consumption and need

There are two approaches to estimating need, rather than consumption, per household.

1   Eliciting the consumer's own perceptions of the amount of fuel required to cook the basic diet and to meet space heating and lighting requirements.
2   Determining the actual energy consumption of households living at, or close to, nationally defined poverty benchmarks in different agro-climatic regions. This will provide an estimate of minimum energy needs.

As populations frequently suffer from 'disguised' shortages, with actual consumption being less than the demand need, it is important to try to establish present and future shortfalls in energy provision. For example, in Tuvalu the demand for LPG outstripped the supply.

Main factors that determine energy consumption

To obtain a complete picture of energy consumption, and to make projections for consumption under a range of different circumstances, it is necessary to understand the factors that determine consumption patterns. However, much less attention is usually paid to the determinants of consumption than to the assessment of consumption itself, although they are equally important for both the projections and assessments of proposed interventions.

Pay particular attention to biomass consumption in urban households, as domestic fuel consumption usually accounts for the greater part of biomass fuel use. But to understand energy consumption by the urban household it is often necessary to differentiate between cities and regions as well. For example, the study by Singh (2013) found there was vast difference between the Funafuti Island and Vaitupu Island communities in Tuvalu in terms of the source of energy used for cooking and lighting. Gas was the major source of energy for cooking (80.6% of the households surveyed) in Vaiaku and Senala villages (Funafuti Island). Kerosene and firewood were used by some households only. In contrast, in Tumaseu village (Vaitupu Island) the main source of energy for cooking was firewood (50%). The remaining households surveyed used gas and kerosene. The main sources of energy for lighting in Funafuti communities (Vaiaku and Senala villages) were electricity, diesel generators and candles. All the Funafuti communities had access to electricity supplied from diesel generators and the electricity supply is 24 hours a day. In Tumaseu village on Vaitupu Island the majority of households surveyed used electricity as their main source of energy for lighting and some households used a domestic diesel generator. However, the electricity supply in Vaitupu Island is for 12–18 hours a day only.

The level of analysis

An initial bioenergy survey will likely be concerned with only those sectors that use the most biomass. These will include villages, urban markets and those industries that use large quantities of biomass energy (e.g. bagasse in the sugar industry, fuelwood for charcoal industries). See Table 5.2.

Analysing domestic consumption

The energy consumption of the household

Household energy consumption is frequently related to household income and size. Although cooking consumes the largest proportion of energy in the domestic sector, there are many other domestic activities that require fuel. These activities may be of considerable importance in a particular locality, but may vary according to the season.

Various possible domestic end uses of fuel are listed below.

Cooking and related

- Domestic food preparation
- Preparation of tea and beverages
- Parboiling rice
- Drying food for storage
- Preparing animal foods

Other regular forms of consumption

- Boiling water for washing
- Boiling/washing clothes

Table 5.2 Levels of analysis of biomass consumption

Sector	Desegregation
Urban household	Income group
Rural household	Income group
Agriculture	Large, small farms
Large-scale industry (commercial)	Food, chemicals, paper, construction, tobacco, beverages, others
Household and small-scale smithing, other	Food, brewing, pottery, industry (informal)
Transport	Air, rail, sea, water, road, private and public vehicles
Commercial	Offices, hotels, restaurants, others
Institutions	Hospitals, schools, armed services, others

- Weaving
- Drying
- Fumigation
- Space heating
- Lighting: domestic and for deterring predators
- Ironing

Occasional forms of consumption

- Food preparation and brewing for ceremonies
- Protection (warding off wild animals, repelling insects, etc.)

Determinants of energy used for cooking

Variation between households in the amount of cooking fuel used occurs for several reasons, including:

- the type of food cooked,
- the method and equipment used for cooking,
- ethnic, class or religious factors,
- the size of the household, and
- income for individuals and the household.

In industrialized countries, fairly standard cooking fuels and equipment are used. However, the specific fuel consumption in developing countries varies considerably (even when the same type of fuel is used) from about 7 to 225 MJ/kg of fuel.

The potential gains from using efficient cooking equipment, e.g. stoves, are enormous. A good stove can save 30–60% in fuel use. Other no less important benefits of stoves include health and hygiene, better cooking environment and improved safety. Tables 5.3 and 5.4 summarize results from Alofa Tuvalu showing variation in selection of preferred cooking fuels and methods.

Table 5.3 Fuel used for cooking in households in Vaitupu, Tuvalu

Fuel/usage	% Respondents
*Use gas at least once per week	59
Use open fire at least once per week	94
Use kerosene every day	100
Use charcoal every day	71
Use open fire every day	59

*Gas use would be more frequent if the supply was reliable.

61% of respondents boiled water before drinking in Vaitupu.

Table 5.4 Fuel used for cooking in households in Funafuti, Tuvalu

Fuel/usage	% Respondents
*Use gas at least once per week	44
Use open fire at least 5× per week	100
Use kerosene every day	100
Use charcoal at least 4× per week	44
Use open fire every day	59

*Gas use would be more frequent if the supply was less expensive.

100% of respondents boiled water before drinking in Funafuti.

Use of coconut husks for boiling water in Vaitupu

Forty one per cent of households questioned in Vaitupu use on average 1.5 kg coconut husks (air dried; the husks from six coconuts) for boiling water each day. This represents an annual use of 0.55 t biomass (or 8.8 GJ) per household which uses an open fire to boil water. Over the total population of Vaitupu, this represents 0.22 t per household per year (equal to the husks from 2.4 coconuts per day).

Total annual consumption of coconut husks for boiling water in Vaitupu is 55 t [887 GJ of useful energy: equivalent to 21 tonnes of oil equivalent (toe); or around 60 t of fuelwood equivalent].

Use of coconut shell charcoal for boiling water in Vaitupu

Forty four per cent of households questioned in Vaitupu use on average 0.2 kg coconut shell charcoal for boiling water each day. This represents an annual use of 8 t (246 GJ or 6 toe) of coconut charcoal in Vaitupu. For those respondents using a charcoal stove, this represents 0.07 t charcoal per year or 1.4 kg charcoal per week for boiling water. Over the total population of Vaitupu, this represents 0.03 t charcoal use per household per year or 0.6 kg per week.

However, production of coconut charcoal has an efficiency of only 15–40%, therefore 26.4 t of coconut shells with an energy value of 529 GJ are required to produce the 8 t of charcoal used annually: 139,187 coconut shells are required each year to produce the charcoal. This is equivalent to 3.5 shells per day for each household with a charcoal stove. Over the total populating of Vaitupu, this represents 1.5 shells per day for boiling drinking water.

Total annual consumption of coconut shell charcoal for boiling water in Vaitupu is 8 t charcoal (246 GJ or 6 toe of useful energy). This requires 26.4 t of coconut shells to produce; 283 GJ (6.7 toe) are wasted in the conversion process.

Communal cooking in Vaitupu

One hundred per cent of respondents had at least one communal cooking activity every 3 months which would involve cooking on an open fire using coconut husk and fuelwood (usually cooking a pig or a large meal).

- Composition of fuel by weight: 21% fuelwood, 7% coconut shell, 72% coconut husk.
- Average volume of material burnt = 0.5 m³.
- Material is not densely packed and mass of material = 32.3 kg (energy value = 17 GJ/t; energy value per fire = 558 MJ).
- Total weight of biomass burnt = 32 t per year (543 GJ, 13 toe or 36 t fuelwood equivalent).

Use of coconut husks for boiling water in Funafuti

Sixty seven per cent of households questioned in Funafuti use coconuts for boiling water at least five times per week. On average 1 kg coconut husks (air dried; the husks from four coconuts) for boiling water each day. This represents an annual use of 0.36 t biomass (or 6 GJ) per household which uses an open fire to boil water. Over the total population of Funafuti, this represents 0.25 t per household per year (equal to the husks from 2.7 coconuts per day).

Total annual consumption of coconut husks for boiling water in Funafuti is 156 t (2,500 GJ useful energy, equivalent to 60 toe).

Use of coconut shell charcoal for boiling water in Funafuti

Sixty four per cent of households questioned in Funafuti use charcoal stoves for boiling water on average 4 days per week, sometimes in conjunction with cooking. On average 0.1 kg coconut shell charcoal is used for boiling water each day. This represents an annual use of 15 t (463 GJ or 11 toe) of coconut charcoal in Funafuti. For those respondents using a charcoal stove this represents 0.04 t charcoal per year or 0.7 kg charcoal per week for boiling water. Over the total population of Funafuti, this represents 0.02 t charcoal use per household per year or 0.4 kg per week.

A total of 50 t of coconut shells with an energy value of 995 GJ (24 toe) are required to produce the 15 t of charcoal used annually; thus, 261,878 coconut shells are required each year to produce the charcoal. This is equivalent to 1.8 shells per day for each household with a charcoal stove. Over the total populating of Funafuti, this represents 1.1 shells per day for boiling drinking water.

Total annual consumption of coconut shell charcoal for boiling water in Funafuti is 15 t charcoal (463 GJ or 11 toe of useful energy). This requires 50 t of coconut shells to produce; 532 GJ (13 toe) are wasted in the conversion process.

Singh (2013) features the case study of Tumaseu village in Vaitupu Island and Vaiaku and Senala village in Funafuti Island (Table 5.5).

The main source of energy for cooking in Tumaseu village is firewood (50%) that is collected (coconut shells, husks and wood from the forest). Some of the households surveyed use gas and kerosene, 27.8 and 22.2% respectively. As 50% of the households earn a living from wages they purchase gas and kerosene for cooking. At the time of the study the average weekly cost for gas was US$29.72 and on average 9 kg of gas is used per week for cooking. The average weekly cost for kerosene among the households surveyed was US$5.12 and on average 2 l of kerosene is used per week for cooking.

Gas is the major source of energy for cooking, being used in 80.6% of the households surveyed (Table 5.6). The average weekly cost for gas was US$27.67 and on average 9 kg of gas is used per week for cooking. Kerosene is used by 9.7% of the households as their main source of energy for cooking. The average weekly cost for kerosene among the households surveyed was US$7.17 and on average 4.5 l of kerosene was used per week for cooking. Firewood is also used for cooking in the communities surveyed: 8.3% of the households collect firewood (such as coconuts branch, shells, husks, breadfruit wood and wood from the forest; on average 60 kilograms of firewood was used per week) and 1.4% of the households buy firewood for household cooking (average weekly cost for firewood was US$6.15).

Firewood that is collected by households (64.7%) is the main source of energy for boiling drinking water in Tumaseu village. Fuelwood is available on Vaitupu Island in abundance as compared to Funafuti Island and so it dominates the source of energy for boiling in Tumaseu village. Gas is the second major source, accounting for 23.5% of the households; however, in Funafuti gas dominated as it is the main source of energy for cooking in Senala and

Table 5.5 Main source of energy for cooking in Tumaseu village, Vaitupu Island

Fuel type	% Respondents
Firewood collected	50
Gas	27.8
Kerosene	22.2

Table 5.6 Main type of energy for cooking in Vaiaku and Senala villages, Funafuti Island

Fuel type	% Respondents
Gas	80.6
Kerosene	9.7
Firewood collected	8.3
Firewood bought	1.4

Vaiaku villages. Electricity supply and kerosene are minor energy sources for boiling drinking water in Tumaseu villages, both showing similar percentage of use that is 5.9%.

Analysing village consumption patterns

A complete village survey on biomass fuel consumption would include all the categories below. For a particular problem, only the relevant (and possible) categories need be selected. Multi-level statistical analysis is a useful tool for analysing many of the relationships described below.

- Pattern of energy use for various fuel-consuming activities in different agricultural seasons, along with the methods of acquiring the fuel (ownership, collection, exchange, purchase, etc.).
- The relationship of the pattern of energy use with family size, land holding, cattle population, income, education, urbanization, etc., so as to predict changes over time.
- Relationships among different categories of households, with regard to energy exchange or ownership of energy assets.
- Possibilities of conversion and fuel substitution under different fuel-price scenarios.
- Assessment of livestock population, forest land, uncultivated waste land, pastures, pattern of crop production, labour, etc.
- Desegregation of biomass energy sources into a number of categories, on the basis of method of acquisition, energy content, moisture content and end use.
- The views of local people (farmers, villagers, women, leaders) as to what they feel are the most important problems being faced, particularly energy shortages/difficulties.
- The source of biomass supplies, and time spent in collection.
- Changes in these patterns of supply within living memory, and the reasons for these changes.

The collection and consumption of biomass fuels can show marked seasonal variation. Surveys should probably be repeated for the different agricultural seasons. This will provide a picture of the different patterns of energy use in relation to the major ways by which biomass is acquired: ownership, barter, free collection and purchase.

Results from Alofa Tuvalu showing consumption of biomass energy for commercial activities on a community scale (Vaitupu) are given below.

Commercial biomass energy use

Kaupule toddy production

Some 48,180 l per year are produced from 30 trees with an average of four taps per tree. This data is included in Table 5.11.

Coconut oil production

Each of the outer islands has a Kaupule coconut oil production facility. Vaitupu also has a second coconut oil mill which is currently not working but is in the process of being renovated.

The Kaupule mill processes around 125 coconuts per day. The price paid to coconut producers is A$0.15 per nut. When the mill is running the average daily production of coconut oil is 13 l. Two people work full time at the mill and receive a wage of A$48.50 each per week, and monthly electricity bills are around A$100. Around 100 kg of coconut husk and shells are burnt each day to dry the copra (see Tables 5.7 and 5.8).

- Energy value of total oil production = 105 GJ (2.5 toe).
- Energy value of total copra production = 185 GJ (4 toe).
- Value added to the process: the crushed copra (copra cake) is sold as a feed for chickens and pigs.

Fuel-using technologies

The technology employed to burn the fuel can have a considerable effect on:

- the amount of fuel required for a particular task,
- the possible complementary end uses.

It may prove important to produce a breakdown of fuel consumption by the technology employed, as was the case with the Alofa Tuvalu survey. The next step is to establish the range of fuels used.

Table 5.7 Coconut oil production on Vaitupu

Total coconut oil (l/year)	Total coconuts used per year	Average nuts per litre coconut oil	Production cost per litre (A$/l)	Copra required (kg/l)	Pith required (kg/l)	Total annual copra production (t/year)
3,120	30,000	9.6	3.32	2.1	3.5	6.6

Table 5.8 Biomass energy required for copra production on Vaitupu

Total biomass burnt (t/year)	Total number of coconut husks used per year	Total number of coconut shells used per year	Energy value of biomass burnt (GJ/year)	toe
24	72,000 (18 t/year)	31,579 (6 t/year)	408	10

Energy sources

A detailed breakdown of the many forms of biomass e.g. twigs, wood, stalls, coconut husk, coconut shell, dung, etc. is often useful, as this enables:

- separation of the needs and problems of different socio-economic groups,
- matching the projections for consumption and supply,
- identification of current and impending fuel shortages.

As the price and availability of fossil fuels are often foremost in the factors affecting biomass consumption, non-biomass energy sources should also be included in the analysis.

To obtain a complete picture of energy consumption, it is necessary to look at all other energy sources that are used in addition to biomass. These can be divided into:

- Animate
- Natural
- Secondary fuels
- Fossil fuels

Animate. Work performed by animals and humans is the fundamental source of power in developing countries for agriculture and small-scale industry. Animals and humans provide pack and draught power, carry head loads, and transport materials by bicycle, boat and cart. Draft animal power should not be seen as a backward technology in conflict with mechanization and modernization. On the contrary, given such a contribution, and the fact that animate power is likely to be the main, feasible and appropriate form of energy for large groups of people for the foreseeable future in many countries, it should be given prime attention. Appendix 4.3 summarizes the main methods of measuring animal and human power.

Natural energy sources. Water, wind and sunlight can all provide alternative sources of energy, and are therefore potential substitutes for bioenergy via waterwheels, transport, hydroelectricity, windmills, solar devices, etc. However, natural energy sources may be only a seasonal asset in some instances.

Secondary fuels. Secondary fuels are sources of energy manufactured from basic fuels. Biogas, ethanol, charcoal and electricity all fall under this heading. Electricity is a good substitute fuel for lighting, but requires expensive technology for other uses. Continuity of supply, and servicing, and spares for equipment are often uncertain and should be considered when looking at secondary fuel use (see Chapter 4).

Fossil fuels

- Coal and coke can relieve demand for biomass fuels, particularly in industrial applications. However, distribution is often affected by dependence on imported petroleum products for vehicles.

- Petroleum products: when imported, the petroleum products kerosene, gasoline, diesel and LPG (i.e. propane and butane) place a heavy demand on foreign exchange. Supply can therefore fluctuate with economic fortunes.
- Kerosene is a much favoured substitute for biomass as a domestic fuel. It has the advantage that the apparatus for its use are fairly cheap.
- Gasoline is the main fuel for personal transport.
- Diesel is the main fuel for freight and public transport, and is used for small-scale generation of electricity, and other uses such as irrigation and water pumping.
- LPG meets the same demand as biogas, and if it is economic and regular in supply it will have an effect on the amount of biomass used for domestic fuel. However, LPG stoves are expensive so its use may be restricted because of this.

By combining fuel-consumption data from the analysis of various end uses and fuel types, it is possible to produce a table/chart giving fuel consumption according to type and sector. Figure 5.1 provides an example of primary energy consumption and Figure 5.2 final energy consumption in Tuvalu.

It is thus now possible to calculate the total consumption of any fuel for a particular end-use. End-use (or final) consumption is, of course, quite distinct from total energy consumption, which includes the energy losses experienced in producing or transporting and energy form.

Changing fuel-consumption patterns

Indicators of changing fuel-consumption patterns

There are five main indicators of changing fuel-consumption patterns:

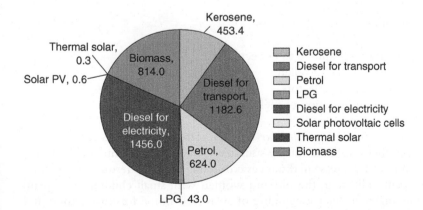

Figure 5.1 Primary energy consumption for Tuvalu, 2005. All figures are given in toe. Solar PV, solar photovoltaic.

Source: Hemstock and Raddane (2005).

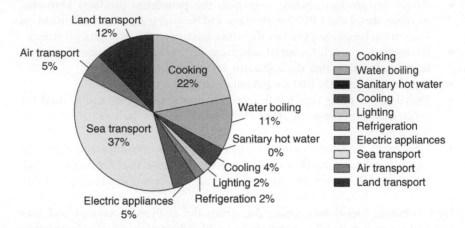

Figure 5.2 Final energy consumption by end use for Tuvalu, 2005 (3,232 toe).

Source: Hemstock and Raddane (2005).

1   fuel collection,
2   type of fuel collected,
3   fuel-using practices,
4   marketing of fuels,
5   fuel supply enhancement.

Monitoring these indicators will give a good idea of changes occurring in consumption patterns.

Fuel collection

Pointers include:

* increase in time required to collect,
* increase in distance travelled,
* change in type of collectors,
* change in transportation of fuel.

Increases in time required and distance travelled. This would necessarily seem to follow from fuel scarcities, but the situation is often more complex than it first appears. While an increase in distances covered is bound to result in an increase in time spent collecting, the old and women with small children will spend more time on intensified scavenging of areas near home for leaves, roots and other low-grade fuels. Resorting to lower grades of fuel will not involve longer journeys, especially when fuel collecting is combined with other activities. In the short term, distances might even be reduced.

Change in who collects fuel. In some cultures, fuel shortages may result in men joining in wood collection, particularly when there is a switch to larger parts of trees, requiring tools not normally available to women. In other cases, children may be withdrawn from school either to collect fuel themselves or to carry out other tasks to give adults more time to fetch fuel.

Change in means of transport. The necessity to travel longer distances may mean that bicycles, handcarts or even draught animals are used to transport the fuel, particularly if men are now involved. This may result in less frequent journeys. However, if it becomes necessary to collect from rough terrain, this may mean a return to head loading.

Change in type of fuel collected

The indicators to look for here are:

- change from dead to green wood,
- change to younger trees,
- change to less preferred parts of trees,
- change to less preferred species,
- change to species with other valuable products,
- change to residues.

Change from dead wood. Switching to green wood, younger trees or less pre-ferred parts or species of trees or plants is a progressive process that will have an increasing effect on fuel-using practices.

Changing the rotation cycle. A shortening of the length of the rotation cycle under shifting cultivation is an indicator of a growing shortage of agricultural land, but will mean a change in fuel consumption to younger, smaller growth.

Switching from fuel species. Changing from fuel species to those with other valuable products is often gender related. Women are initially reluctant to take fuelwood from trees that provide fruits and other foods, medicines and craft-making material. Men will wish to reserve those species which generate a cash income or furnish implements and tools.

Changing to agricultural and animal residues. High-grade crop residues are always popular fuels, as they are generally easy to obtain and store. However, when fuel is scarce people may resort to low-grade residues and dung. The use of these low-grade fuels sometimes carries a social stigma.

Fuel-using practices

Changing fuel-using practices are observed through:

- an increase in duration of cooking times,
- increase in intensity of cooking,
- introduction of fuel-saving devices,

- reduction in fuel-using activities,
- change to consumption patterns using less, or more, fuel.

An increase in the duration or intensity of cooking, and the introduction of fuel saving devices. When only low-quality fuel is available there is an increase in the time required for cooking. Where household income allows, there may be a switch to stoves with a greater efficiency, rather than devoting more time to collecting fuel.

Reduction in the amount of fuel-using activities. This may involve a cutback in household industry. However, before income generating activities are curtailed, there may be switches to foods requiring less cooking, or fewer cooked meals may be eaten.

Change to consumption patterns using less, or more, fuel. The introduction of a substitute fuel such as kerosene will reduce overall consumption, but the need to generate income to pay for it might drive a lower-income household into starting an enterprise such as brewing, which would raise their overall consumption of fuelwood.

Marketing of fuels

Market indicators include:

- an increase in the range of fuels bought and sold,
- an increase of marketed fuels in total consumption,
- an increase in the cost of individual fuels,
- an increase in household expenditure on fuel.

An increase in the range of fuels commercially transacted. The development of a market economy will change consumption both by the import of substitute fuels and the export of charcoal and fuelwood.

An increase in the proportion of marketed fuels in the total consumption. In some areas this will denote a reduction in the availability of biomass fuel, but in others it is a sign of increasing urbanization and prosperity, e.g. as often the case with charcoal.

An increase in the cost of individual fuels. The price of substitute fuels is so tied up with national and international economics that an increase in the price of a fuel such as kerosene does not usually indicate an increase in demand. However, an increase in the cost of kerosene will usually lead to an increase in the consumption of biomass fuels by the many households where kerosene is only marginally affordable. It may also lead to a price increase in other substitute fuels.

An increase in the proportion of household expenditure devoted to fuel. Before this is used as an indicator of the increasing use of commercialized fuels the price trends of commercial fuels must be carefully considered.

Supply enhancement

This is recognized by:

- change of cropping pattern,
- increase in planting of fuel crops.

Change of cropping pattern. Most changes in cropping (or alterations in the number of livestock, or the abandonment of shifting agriculture for settled occupation) will result in differing amounts of residue being available as fuel. If crop residues constitute a major part of the existing fuel supply, then any change in cropping patterns will effect fuel availability and hence consumption. However, the significance of residues in the household economy may be reduced if new farming ventures provide an increase in income.

Increased planting of fuel crops. Fuel crops are often planted as cash crops for sale in urban areas. Therefore, the appearance of fuel crops does not necessarily imply a local change in fuel-consumption patterns, although the loss of residues from crops replaced by trees may affect lower income households. On the other hand, fuelwood demand is increasingly being met from forest stocks, which are decreasing at an alarming rate in some countries.

Variations in fuel consumption

Variations between localities

There are often large variations in biomass consumption between neighbouring localities. To capture these variations it is often desirable to select clusters of communities from several regions. The extent to which activities vary between communities will determine the number and distribution of locations covered. Several factors can cause variations between neighbouring localities, including:

- different types of food cooked,
- different methods or means of cooking (baking instead of boiling, open fire or stove),
- different practices associated with different ethnic groups, economic classes and religions,
- different income groups: disaggregation by income group is generally useful for urban consumers. For example, those less well-off are almost exclusively dependent on biomass, middle-income households use charcoal, kerosene, gas or electricity, and high-income households use combinations of charcoal, kerosene, gas, electricity or wood depending on what is being cooked and the relative prices of the different fuels.

Variation of fuel consumption over time

Variations in fuel consumption are either short term, or seasonal and annual.

Short-term variation. Fuel use recorded over a number of consecutive days can sometimes show large fluctuations. There are several factors that may explain this variation.

- The amount of fuel required from day to day may vary for the following reasons.

  o   Some fuel-using activities are only undertaken on certain days. Brewing is not a daily occupation, and food for sale is more likely prepared for local market days. Many household industries require fuel intermittently, e.g. steeping woven cloth, or firing pots.

  o   The number of guests and the members of the household eating elsewhere may vary considerably from day to day. Although the overall exchange in hospitality may equalize fuel consumption over time, the effect on day to day consumption may be large.

  o   Similarly, the number of hired employees fed (or household members employed and fed elsewhere) can vary daily. The irregular nature of household industries and the interchange of employer/employee roles between neighbours with differing enterprises will alter daily rates of consumption. Larger agricultural employers are likely to hire labour intermittently.

- Changes in the behaviour of the consumer.

  o   A change of user can cause a considerable change in consumption, depending on the competence of the new user. This is particularly important when an open fire is used and the quality of the fuel is variable.

  o   The amount of time the user has at his or her disposal. Pressure from other commitments may lead to shortage of time for collecting or economic use of fuel, and may give rise to excessive consumption.

- Other factors for consideration.

  o   Weather conditions can make important differences in consumption. When the fire is exposed, windy conditions will make the fuel burn quickly. Unseasonable temperature changes may need additional space heating, and unexpected rain storms will result in damp wood with a poor performance, and difficulties in collection and storage.

These short-term variations are clearly of critical importance for the interpretation of any physical measurements of consumption that are made. But even where data is collected verbally, variations may mean that the concept of a 'normal' level of consumption is relatively meaningless.

Seasonal or annual variations mean that it is important for proper assessment of biomass fuel use over the whole year. Fuel consumption may also vary to a greater or lesser degree on a seasonal basis, or from year to year. Some of the causes of this variation are listed below.

Seasonal variation

- The amount of fuel required may vary through seasonal change in fuel-consuming activities.

  o The composition of the household may vary with the seasonal demands for agricultural labour. Extra workers may be hired. Family members may get local employment, or migrate seasonally to find work. Small-scale industries such as brick-making or tobacco curing are also seasonal, and may require hired labour.

  o The number of ceremonies varies throughout the year. Social or religious gatherings and ceremonies do not occur at regular intervals through the year, and may distort seasonal fuel consumption.

  o Availability of food: more food is usually available after the harvest than before it.

  o The portions or the diet eaten raw or cooked may vary, together with the quantities of food available for consumption.

  o Temperature and the need for surface heating. At higher altitudes or latitudes extra fuel and food is consumed for space heating during the cold season.

- The availability of fuel varies with the seasons.

  o Crops providing residues for fuel have specific harvest times. Periodic activities that require fuel are often undertaken after the harvest when crop residues are abundant, particularly if storage is difficult, or labour is available.

  o The weather affects the ease with which fuel is gathered. Except where large-scale storage is available, rainy seasons affect both the quantity of fuel it is possible to collect and its burning qualities. Long periods of windy weather also lead to extra fuel consumption.

  o Agriculture and other labour-intensive activities affect the time available for collecting fuel. Demand for labour, particularly women's labour, is greatest at sowing and harvest time. Time available for collecting fuel and cooking is thus reduced. In the dry season, the need to fetch water from greater distances may also compete with fuel collection.

  o Storage of fuel can be a problem. Access to storage facilities may depend on wealth and status. High income households have the space and capital to build stores and to pay labour to fill them. Those with lower incomes usually do not have large quantities of crop residues to store. Different types of storage will vary in their vulnerability to damp. Some woods do not last as long in store as others, and may be more susceptible to insect attack.

Annual variations

- Annual weather variations may affect the growth of plants of all types, and therefore the amount of residue (and to a lesser extent wood) available must influence the rate of regeneration of fuelwood supplies.

- Differing trends in market prices will determine the crops grown and the residues available. Timber may be grown to satisfy urban fuel demands, but this is likely to diminish rather that increase fuel supplies. Increase in the price of fossil fuels will transfer both urban and rural demand to fuelwood. Conversely, a decrease in the cost of fossil fuels may reduce the demand for fuelwood. However, once a charcoal supply chain is established it is likely to remain fully active.

Projecting supply and demand

In order to make meaningful projections of supply and demand data, must usually reflect total above-ground woody biomass present in an area that could be used as a source of energy as well as present consumption. It should be borne in mind that woody biomass has many other applications with an even higher value than fuel, such as timber, plywood, pulpwood, etc. Thus, only a small proportion, e.g. branches, tops, unmerchantable species, etc., might end up as fuelwood. The initial end result of any biomass inventory should be a database constructed from tally sheets for each of the samples, e.g. number of trees or bush species with measurements such as stem diameter, crown diameter, etc. Measurement of parameters can be used in combination with other data, e.g. tree-weight tables to determine the standing stock of woody biomass. The result should be a database that shows for each strata or vegetation type the weight of woody biomass per unit area by size classes. Sustainability of the biomass resource can then be assessed by measuring consumption against availability.

There are various methods for predicting supply and demand: (1) constant-trend based projections, (2) projections with adjusted demand, (3) projections with increased supplies, (4) projections including agricultural land and (5) projections including farm trees. From the point of view of a project planner it may be wise to use more than one method of projecting supply and demand.

Constant-trend based projections

These assume that consumption and demand grows in line with population growth and that there is no increase in supply. It is a useful way to identify any resource problems and possible actions to bring supply and demand into a sustainable balance. Essentially, consumption grows with population growth, and supplies are obtained from the annual wood growth and clear felling of an initially fixed stock of trees. However, as wood resources decline, costs will increase and consumption will be reduced by fuel economics and substitution of other fuels.

Projections with adjusted demand

These are a useful step to examine reductions in per capita demand and its effects in declining wood resources. The adjustments can then be related to policy targets, such as improved stove programmes or fuel substitution.

Projections with increased supplies

Wood supplies can be increased by a variety of measures, e.g. better management of forests, better use of wastes, planting, use of alternative sources such agricultural residues, etc. Targets for these additional supply options can easily be set by estimating the gap between projected woodfuel demand and supplies.

Projections, including agricultural land

In most developing countries the spread of arable and grazing land, together with commercial logging in some areas, is a major cause of tree loss. When land is cleared by felling and burning, the result is greater pressure on existing forest stocks for fuelwood. If the wood cleared is used for fuel this will contribute to release this pressure.

Projections including farm trees

Trees have multiple uses, e.g. fruit, forage, timber, shelter, fuelwood, etc. These farm trees, which are fully accessible to the local consumers, are often a major source of fuel in many rural areas and hence should be included in any projection models.

Alofa Tuvalu survey results for coconut production and use in Tuvalu

Coconuts used for feeding humans and pigs

- Number of coconuts used for feeding pigs: Table 5.9
- Number of coconuts used for feeding people: Table 5.10.
- Toddy production: Table 5.11.

Commercial biomass energy use

All outer islands have a small coconut oil production facility like the one detailed for Vaitupu (Table 5.12).

Table 5.9 Coconut pith fed to the pigs in Tuvalu

Total number of pigs*	Total number of coconuts used per day for feeding pigs in Tuvalu	Number of coconuts used per year for feeding pigs	Energy content of pith fed to pigs (GJ/year)
12,328	13,534	4,940,266	25,571 (608 toe)

*Using the Alofa Tuvalu estimate for the number of pigs in Tuvalu.

Table 5.10 Coconuts used for human consumption in Tuvalu

Total number of people	Total number of coconuts used per day for feeding people in Tuvalu	Number of coconuts used per year for feeding people	Energy content of pith fed to people (GJ/year)
9,561	2,930	1,534,094	7,941 (189 toe)

Table 5.11 Toddy production in Tuvalu

Total number of toddy trees in Tuvalu	Total number of toddy taps in Tuvalu	Estimated total toddy production (litres per year)
1,636	4,931	1,979,818

Table 5.12 Coconut oil production*

Total coconut oil (l/year)	Total coconuts used per year	Average nuts per litre coconut oil	Production cost per litre (A$/l)	Copra required (kg/l)	Pith required (kg/l)	Total annual copra production (t/year)
16,800	198,000	10	3.92	2.2	3.6	43.56

*Energy value of total oil production = 671 GJ (16 toe).

Table 5.13 Biomass energy required for copra production*†

Total biomass burnt (t/year)	Total number of coconut husks used per year	Total number of coconut shells used per year	Energy value of biomass burnt (GJ/year)	toe
192	576,000	252,632	3,264	78

*Energy value of total copra production = 1,220 GJ (29 toe).

†Value added to the process: the crushed copra (copra cake) is sold as a feed for chickens and pigs.

Wasted energy and total biomass energy use

- The total number of coconuts used (to feed to pigs and people) is 6,672,361 (energy available from husk and shells: 52,044 GJ or 1,239 toe).
- The total number of coconut husks used as a fuel (domestic and commercial) = 5,136,242 (with an energy value of 20,545 GJ or 498 toe).
- The total number of coconut shells used as a fuel (domestic and commercial) = 2,965,101 (with an energy value of 11,267 GJ or 268 toe).

- Total biomass energy consumption (domestic) = 31,350 GJ (746 toe).
  (useful energy after some conversion to charcoal = 26,112 GJ or 622 toe).
- Total biomass energy consumption (commercial) = 2,856 GJ (68 toe).

Estimate of coconut production in Tuvalu

Coconut production in Tuvalu is enough to meet the requirements for domestic biomass energy, current indigenous food use and 56% of current fuel use of the outer island boats. The boats use half the total fuel oil imported into Tuvalu. Therefore, current coconut productivity in Tuvalu could easily replace 25% of current total fuel oil use. Payment to copra growers would total A$575,093 annually at prices currently agreed with producers on Vaitupu (Table 5.17).

As stated, two aims of the Alofa Tuvalu were to assess the amount of biomass energy resource available for biogas digestion and coconut oil biodiesel production projects, and to ensure that they were sustainable in terms of current biomass production.

Table 5.14 Wasted energy from coconuts

Total number of shells unused each year	Energy value of unused shells (GJ/ year)	Energy value of shells (toe)	Total number of husks unused each year	Energy value of unused husks (GJ/ year)	Energy value of husks (toe)	Total energy available from unused husks and shells (GJ/ year)*
3,707,260	14,088	335	1,536,118	6,144	146	20,232 (482 toe)

*The unused husks and shells (which total 482 toe) would be an ideal fuel for a series of gasifiers for electricity generation for each island's respective grid.

Table 5.15 Wasted energy from pigs

Total number of pigs in Tuvalu (head)	Total annual production of pig manure (t/year)	Total annual production of energy from pig manure (GJ/year)	Total amount of manure available annually for use in biogas digester* (t/year)	Total amount of energy available annually for use in biogas digester* (GJ/year)
12,328 (60% = 7,397)	3,600	32,400 (771 toe)	2,106	19,440 (463 toe)

*Assumes 60% collection efficiency as some waste is used for compost and some will be difficult to collect.

Table 5.16 Coconut production in Tuvalu

Total number of hectares under coconut	Estimate of total number of productive trees	Estimate of total number of trees over 60 years	Total coconut production (nuts per year)	Number of hectares requiring re-planting	Theoretical availability of coconuts for biomass energy production (nuts per year)*
1,524	267,760	68,929	14,141,100	391	7,468,739

Source: Based on McLean et al. (1986); Seluka et al. (1998); Hemstock and Raddane, 2006.

*Total production minus use.

Table 5.17 Theoretical production of coconut oil from unused coconuts in Tuvalu

Total number of litres of oil per year (l/year)	Energy value of coconut oil (GJ/year)	toe of oil produced	Copra production (t/year)*	toe of the copra produced	% of boat fuel replaced by coconut oil
777,184	26,168	623	1,643	39	56

*This estimate is based on coconut production per productive tree in Tuvalu and is more conservative than the figure for copra production of 1.2 t/ha used by Trewren (1984). However, it is less than Trewren predicted if you use figures of 215 trees/ha, 60 nuts/tree and 0.187 kg copra/nut.

Conclusions

As emphasized previously, biomass energy comprises many components – such as production, conversion, use and conservation – each further subdivided into many equally important subcomponents. Accurate consumption and productivity assessments are a fundamental tool for project planning and subsequent management of biomass resource development and conservation programmes. Without reliable data for both consumption and availability it is difficult to undertake meaningful policy and planning decisions. Many decisions in the past have been made from a very poor footing which has resulted in many project failures.

Notes

1 French based international NGO. Alofa Tuvalu, 30 rue Philippe Hecht, 75019 Paris, France.
2 Cooking and boiling water were assessed separately in the Alofa Tuvalu survey and as a result the introduction of solar water heaters for providing sanitary and drinking water has been recommended.

References and further reading

ETC (1990) Biomass assessment in Africa. ETC (UK) in collaboration with Newcastle Polytechnic and Reading University. World Bank, Washington DC.

ETC (1993) *Estimating woody biomass in Sub-Saharan Africa*. ETC (UK) in collaboration with Newcastle Polytechnic and Reading University. World Bank, Washington DC.

GOVERNMENT OF ZIMBABWE (1985) Geographical extent of vegetation types, estimates of total and accessible areas surviving in 1984 and their growing stock increment. In *Biomass assessment: woody biomass in the SADCC region* (1989) Millington, A. et al. (eds). Earthscan Publications, London.

HALL, D.O. AND HEMSTOCK, S.L. (1996) Biomass energy flows in Kenya and Zimbabwe: indicators of CO_2 mitigation strategies. *The Environmental Professional* 18, 69–79.

HEMSTOCK, S.L. (2005) *Biomass energy potential in Tuvalu (Alofa Tuvalu)*. Government of Tuvalu Report.

HEMSTOCK, S.L. (2013) The potential of coconut toddy for use as a feedstock for bio-ethanol production in Tuvalu. *Biomass and Bioenergy* 49, 323–332.

HEMSTOCK, S.L. AND HALL, D.O. (1995) Biomass energy flows in Zimbabwe. *Biomass and Bioenergy* 8, 151–173.

HEMSTOCK, S.L. AND RADDANE, P. (2006) *Tuvalu renewable energy study: current energy use and potential for renewables*. Alofa Tuvalu, French Agency for Environment and Energy Management (ADEME). Government of Tuvalu.

HOSIER, R.H., KATARERE, Y., MUNASIREI, D.K., NKOMO, J.C., RAM, B.J. AND ROBINSON, P.B. (1986) *Zimbabwe: energy planning for national development*. Beijer Institute, Stockholm.

MCLEAN, R.F., HOLTHUS, P.F., HOSKING, P.L., WOODROFFE, C.D. AND KELLY, J. (1986) *Tuvalu land resources survey, vol. 1–9*. UNDP/FAO Contract DP/TU/80/001-1/AGOF. University of Auckland, New Zealand.

MILLINGTON, A., TOWNSEND, J., KENNEDY, P., SAULL, R., PRINCE, S. AND MADAMS, R. (1989) *Biomass assessment: woody biomass in the SADCC region*. Earthscan Publications, London.

SELUKA, S., PANAPA, T., MALUOFENU, S., SAMISONI AND TEBANO, T. (1998) *A preliminary listing of Tuvalu plants, fishes, birds and insects*. The Atoll Research Programme, University of the South Pacific, Tarawa, Kiribati.

SINGH, R.D. (2013) *Using, managing and controlling energy sources for sustainable development: a gender analysis across communities in Vanuatu and Tuvalu*. Masters of Science in Climate Change thesis, Pacific Centre for Environment and Sustainable Development, University of the South Pacific, Laucala, Fiji.

TREWREN, K. (1984) *Coconut development in Tuvalu*. Ministry of Commerce and Natural Resources, Funafuti, Tuvalu.

Statistics bibliography

http://duke.usask.ca/~rbaker/stats.html (basic statistics)

http://obelia.jde.aca.mmu.ac (statistics and survey design information)

http://www2.chass.ncsu.edu/garson/pa765/statnote.htm (on-line stats text book)

http://www.pp.rhul.ac.uk/~cowan/stat_course_01.html (more stats)

BETHEL, J. (1989) Sample allocation in multivariate surveys. *Survey Methodology* 15, 47–57.

BRAITHWAITE, V. (1994) Beyond Rokeach's equality-freedom model: two dimensional values in a one dimensional world. *Journal of Social Issues* 50, 67–94.

FELDT, L. AND BRENNAN, R. (1989) Reliability. In *Educational measurement*, Linn, R. (ed.). Macmillian Publishing Company, Basingstoke, pp. 105–146.

GIBBINS, K. AND WALKER, I. (1993) Multiple interpretations of the Rokeach value survey. *Journal of Social Psychology* 133, 797–805.

GOLDSTEIN, H. (1995) *Multilevel statistical models.* Halstead Press, New York.
LONGFORD, N. (1993) *Random coefficient models.* Clarendon Press, Oxford.
VALLIANT, R. AND GENTLE, J. (1997) An application of mathematical programming to a sample allocation problem. *Computational Statistics and Data Analysis* 25, 337–360.

Appendix 5.1   House survey

Socio-economic characteristics

1   **Are you:**   O Male   O Female

2   **What was your place of birth?**_____

3   **Please indicate the year you were born:**_____

4   **How long have you lived in this village?** _____

5   **What is the annual income for this household?**_____

6   **What has been the main source of income for this household over the last 12 months (select only one)?**

   O No income   O Wages/salary   O Own business   O Sale of products (fish, handicraft, crops, etc)

   O Land lease   O House rent   O Remittances   O Other source (specify)
   _____

7   **Who manages the household income?**_____

8   **What is your highest level of education (select only one)?**

   O No school completed   O Primary school   O Secondary school

   O Diploma   O Degree   O Masters   O PhD

9   **How many people live in this household?**   Adults:_____   Children: _____

10   **What is your <u>major household expense</u> (select <u>only one</u>)?**

O Food   O Water   O Energy

O Transport   O Healthcare   O Education   O Other (specify) _____

Energy resources: access and management

11   **Does this <u>household</u> have any of the <u>following goods</u> (select <u>all that apply</u>)?:**

O Boat   O Hot water system   O Bath or shower   O Motor vehicle

O Refrigerator O Washing machine   O Television   O Video/DVD
                                                          player

O Telephone   O Mobile telephone   O Computer   O Internet
  landline

12   **What is the <u>main source of lighting</u> for this household (select <u>only one</u>)?:**

O Electricity   O Diesel   O Kerosene lamps   O Other (specify)
  supply   generator
                                                       _____

O Solar   O Candles   O Benzene lamps

13   **Approximately how much does this <u>lighting cost you a week</u>?**

	Electricity	Generator	Kerosene lamps	Solar	Candles	Benzene	Other____
Cost ($)							

14   **Who <u>manages</u> these sources of <u>lighting</u>?** _____

15   **What is the <u>main type of energy for cooking</u> for this household (select <u>only one</u>)?:**

O Electricity supply   O Gas   O Kerosene

O Firewood collected   O Firewood bought   O Other (specify) _____

16   **Approximately how much does this <u>energy for cooking cost you a week
     and what is the amount</u>?**

	Electricity	Gas	Kerosene	Firewood (collected)	Firewood (bought)	Other _____
Cost ($)						
Amount (weight, litres number of bottles)						

17 **Who** does most of the **cooking?**_____

18 **How long** does this person spend **cooking every day?**_____

19 **Who manages** these sources of **energy for cooking?**_____

20 If you use **firewood for cooking**, what is the **main source** (select **only one**)?

    O Mangroves   O Forest   O Dead plant material   O Charcoal

    O Coconut shells   O Coconut husks   O Wood   O Other_____

21 If you use **wood or forest materials, which tree species?**_____

22 **Who** collects the **firewood?**_____

23 **How long** does this person take to collect **firewood per week?**_____

24 **Do you use:** O Wet firewood O Dry firewood

25 If an **open fire** is used for cooking, where is it **located?**

    O Inside (closed space)   O Outside (open air)

26 Do you **boil drinking water?** O Yes O No

27 **How** do you boil your water?

    O Electricity supply   O Gas   O Kerosene

    O Firewood collected   O Firewood bought   O Other (specify) _____

28 What **type** of stove/**oven** do you **own?** O Gas O Kerosene O Firewood

29 **If firewood is used, what type of stove is it?**

    ○ Hearth oven   ○ Three stone   ○ Pykie (Three-legged Pot)
                                 cooking fire

    ○ Oven with chimney   ○ Other type of oven   ○ Charcoal stove

30 **How many times a week do you cook or boil water with firewood?**_____

Issues of concern

31 **Please indicate your level of concern regarding each of the following issues in your local community:**

Issues in your community	No concern	Slight concern	Some Concern	High concern	Extreme concern
Availability of and **access** to **freshwater**					
Decision-making over **freshwater** supplies					
Availability of and **access** to energy sources for **cooking** and **lighting**					
Cost of energy for **cooking** and **lighting**					
Decision-making over **energy** supplies					
Natural disasters (floods, drought)					
Climate change					
Availability of and **access** to **transport**					
Protection of the local **environment**					

32 **If you have any other comments, please write them here:**

_____
_____
_____
_____

Many thanks for your time in completing this survey.

Appendix 5.2   Household fuelwood inventory

	Quantity (for one household meal)	What (sample wood species name; plus other types used)	Weight (total)	Size (one piece)	Number of people for an average household meal	Moisture content	Photo
Household 1							
Household 2							
Household 3							
Household 4							
Household 5 etc.							

Quadrant survey

Species name	Number (tally)	Height (m)	Width (circumference) (cm) (DBH calculated later)

6 Assessing sustainability for biomass energy production and use

Rocio Diaz-Chavez

Introduction

The extensive literature on sustainability demonstrates the idea of expanding the concept to a more integrated, participatory and anticipatory decision-making process. Although the main ideas started in the 1970s as part of the concerns on depletion of resources and growth limits it was not until 1987 with the Brundtland Report that sustainability acquired global recognition (WECD, 1987). The basic idea was to link environment and development to identify problems but also solutions. This was followed by the international conferences from the United Nations (UN) in Rio de Janeiro in 1992 and 10 years later in 2002 (WSDD, 2002). Despite many differences, there are some basic characteristics that can be recognized globally. Gibson (2005) provides a review on the concept of sustainability with some of these characteristics being common for all. Of the basic concepts probably the most recognizable include that the concept refers to short- and long-term well-being, that it covers all main issues on decision making and that the concept is universal and context-dependent.

According to Gibson (2005), environmental assessment refers to the processes designed to further environmental issues in the planning and implementation of different activities. It has taken the form of different tools, including Environmental Impact Assessment (EIA) or social and ecological assessment. This has been the leading approach for around 30 years. For several years the debate has been about putting sustainability assessment at the front of the analysis of the viability of supply chains. Several issues, such as greenhouse gases (GHG), land use and biodiversity, have been required in some policy and legal instruments, e.g. the EC Renewable Energy Directive (EC, 2009). Social aspects, on the other hand, have been taken into account in more depth by non-governmental organizations (NGOs) and sustainability standards. Furthermore, new issues are considered now to differentiate between biomass feedstocks (e.g. agricultural crops, dedicated crops, lignocellulosic residues) and the new uses for a green economy, e.g. biorefineries.

This chapter reviews selected tools and methodologies that can be used to assess and identify whether the different practices for biomass production and use for energy can be deemed as sustainable. Practical tools such as Environmental

Impact Assessment (EIA), Social Impact Assessment (SIA) and Sustainability Assessment (SA), and more recently sustainability reporting, are explored along with examples. Appendix 6.1 summarizes examples of methodologies.

Sustainable development and methodologies for assessing biomass for energy

SA has evolved from the separate assessment of ecological, social and economic issues to an integrated approach where governance also forms part of it (Gibson, 2005; Diaz-Chavez, 2011). According to Gibson (2005) assessment processes need to be designed within an iterative system defined with clear objectives, evaluation and decision criteria. In this form there can be real integration and a move away from traditional pillar thinking (Sheate, 2012).

Environmental assessment and management have improved through the use of different tools and methodologies. They are varied and originate from different disciplines or knowledge and applicability areas. Table 6.1 presents examples of these tools and where they have been traditionally allocated in either 'knowledge areas' or applicability. The list is not exhaustive and currently the application is not limited to one single methodology or tool in one plan or project.

These tools, methodologies and frameworks have been applied to a variety of planning and implemented projects and their use has reached the bioenergy sector. There have been reviews of the different methodologies and tools in use or with the potential to assess projects in bioenergy. Some of these methodologies are explained below.

The Global Bioenergy Partnership (GBEP) published a report in 2010 with a review of analytical tools. The report presents a range of tools mainly based on multiple criteria assessment and focused on its application by decision makers in different regions of the world. The first part of the report focuses on the analytical tools. They are classified into three groups:

Table 6.1 Examples of assessment tools and methodologies

Environmental	Engineering	Social and economic	Business	Integrated
Environmental Impact Assessment	Life-Cycle Assessment	Social Impact Assessment	Environmental Management Systems	Sustainability Assessment
Carbon footprinting	Risk assessment	Health Impact Assessment	Corporate Social Responsibility	Integrated Impact Assessment
Ecological footprinting	Material flow assessment	Cost-benefit analysis		Impact Assessment
Land use modelling		Economic modelling		Strategic Impact Assessment

1   *spatial planning for bioenergy production*, for land use analysis (e.g. Geographic Information Systems, GIS),
2   *technology options and potential*, for techno-economic analysis,
3   *implementation options and impacts*, for integrated assessment and associated impacts of bioenergy.

The second part of the report contains technical back-up information, intended to provide decision makers with technical resource material ranging from stakeholder engagement to mitigation options and good practices. This information covers topics such as air, water, ecosystems and biodiversity and some social issues such as land tenure and income generation (GBEP, 2010).

Another compilation produced by the Food and Agriculture Organization (FAO) was through the Bioenergy and Food Security Criteria and Indicators (BEFSCI) project, which provided a review of 30 relevant tools and methodologies related to the bioenergy sector (Beall et al., 2012). These tools were selected to assess *ex ante* and *ex post* individual projects or for the whole sector in a region. The tools focused on environmental and social impacts rather than on the whole sustainability assessment. They are grouped under three main knowledge areas: (1) *environmental* (biodiversity, water, soil, GHG), (2) *social* (food security, community development, gender equity, energy security and local access to energy), (3) *cross-cutting issues* (land, economy, technology, trade, economy). Each topic provides a range of tools and a description regarding its application (e.g. planning or monitoring, government or operator). Tools range from descriptive appraisals to models.

Traditional environmental management tools have also been used to assess projects in the bioenergy sector. EIA is a valuable tool with legal requirements in many countries around the world. The assessment reports can provide insight in how countries manage sustainability challenges. It is important to note that in some countries EIA is normally applicable to *ex ante* projects that are considered (through a screening process) to have negative impacts on landscape and other aspects. Agriculture is normally not considered as a change in landscape. In fact, in the UK the EIA (Agriculture) (England) (No.2) Regulations 2006 Act (Natural England, 2014) 'protects uncultivated land and semi-natural areas from being damaged by agricultural work, and guards against possible negative environmental effects from the restructuring of rural land holdings'.

The Roundtable on Sustainable Biomaterials (RSB) also includes in its standard a principle and criteria that requires an Environmental and Social Impact Assessment (ESIA).The criterion reads: 'Criterion 2a. Biofuel operations shall undertake an impact assessment process to assess impacts and risks and ensure sustainability through the development of effective and efficient implementation, mitigation, monitoring and evaluation plans' (RSB, 2011). This can be produced following the guidelines from the RSB (RSB-GUI-01-002-01) or those presented as a legal requirement in the country where the economic operator will be working.

Englund et al. (2011) presented a review of EIA selection for biofuels projects. The review covered projects for biofuels production with the aim for exporting to the EU. Therefore the analysis of the EIAs was done following the Renewable Energy Directive sustainability criteria (EC, 2009). The authors reviewed 19 statements in three different regions (Asia, Latin America and Africa). The projects also covered different feedstocks and mills: palm oil, sugar cane, jatropha, eucalyptus plantations, ethanol and biodiesel plants. The analysis of the statements reviewed some of the main impacts identifiable in bioenergy and biofuel projects under environmental and social topics as well as GHG. The report concluded that even though EIA legislation exists, it is insufficient from a biofuels perspective. As the concept of 'EIA' seems to be familiar to the decision makers it might make an improvement of EIA legislation easier to realise. Nevertheless, EIA legislation is not sufficient without enforcement. One of the main limitations of the analysis produced by Englund et al. (2011) is that the review was done only on 19 reports that were publicly available online and were restricted by language, e.g. only those in English were reviewed (Englund, personal communication).

Strategic Environmental Assessment (SEA) and Strategic Impact Assessment (some consider these terms synonymous because the environment also considers social and economic issues) are also useful tools if applied. SEA can be applied to assess policy, plans and programmes but also can be used for a sector (e.g. bioenergy) or a region. The Organization for Economic Cooperation and Development (OECD) and the Swedish Development Agency (SIDA) published a guideline on SEA and biofuel development (OECD, 2011). The report provides guidelines on the main steps of SEA: (1) establishing the context for the SEA, (2) implementing the SEA, (3) informing and influencing decision making and (4) monitoring and evaluation. Although some countries do not yet implement SEA (and particularly in this sector) the government of Kenya followed a very similar methodology to SEA to write and implement their national biofuels policy. The government of Tanzania is finalizing a state-of-the-art national biofuels policy. SEA offers a great advantage, especially when addressing cumulative impacts (Diaz-Chavez, 2011).

The use of Life-Cycle Assessment (LCA) has also contributed to sustainability assessment of the bioenergy sector and provides a methodology that is standardized and considers the whole supply chain rather than just the production of the feedstock. Black et al. (2011) reported that there are over 250 different scenarios for bioenergy production systems using commodity crops. They also presented a review on the concept of biofuel GHG and sustainability metrics with examples from the UK. They assessed willow in a lignocellulose ethanol production system to demonstrate how GHG emission outcomes can be reviewed for 'new' crops and technologies. The analysis of the supply chain LCA for biofuel production described a biofuel supply chain for lignocellulosic ethanol using an LCA modular approach. The main issue is in designing the LCA (e.g. definition of systems boundaries, allocation of impacts and choice of data sources), availability of good-quality data, loss of resolution in 'simplified' models and rebound effects. The authors concluded that there are a range of criteria which must be considered at the local, national and global levels when assessing supply chains based on biomass feedstocks for bioenergy provision. Methodologies to address LCA and sustainability have advanced but data availability still remains an issue for certain feedstocks and impacts.

The EU published a report providing guidance for detailed LCA studies and provides the technical basis to derive product-specific criteria, guides and simplified tools (EC, 2010). LCA has traditionally been used for environmental impacts (Environmental LCA, E-LCA) but it has been extended to consider a full sustainability assessment (Sustainability LCA, S-LCA). This integration is possible because the basis of any LCA should be the technical life-cycle model of the analysed product that is the full supply chain. The sustainability LCA also includes Life-Cycle Costing (LCC) and social LCA (sLCA) and has been deemed to play a crucial role in moving towards sustainable consumption and production (Wolf et al., 2012; Valdivia et al., 2013). sLCA is the most recent methodology that covers some social aspects (e.g. job creation, equal pay for women, stakeholder participation) (UNEP, 2009).

The use of social and socio-economic life cycles can provide additional information to the impact analysis (UNEP, 2009). This applies not only to the product but also to green procurement. Whereas an E-LCA will mainly focus on collecting information, mostly physical quantities related to the product and its production/use and disposal, a sLCA will collect additional information on organization-related aspects along the chain (Diaz-Chavez, 2012a). Figure 6.1 illustrates the specificities of the techniques.

sLCA is still a novel methodology and there are only a few reports using it. Although the main benefit is to include life-cycle thinking on social issues in the supply chain, it is still limited. To address social and economic impacts it is suggested to use a combination of SIA, sLCA and SA to link it with the environmental assessment (Diaz-Chavez, 2012a). A novelty of the sLCA is the possibility of using the hotspots database where integrated data on social issues can be accessed (SHD, 2015). Appendix 6.1 shows an example of the combined methodology and assessment.

Regarding land use and land use change, the methodologies have normally focused on land modelling. Models have also been developed to address the

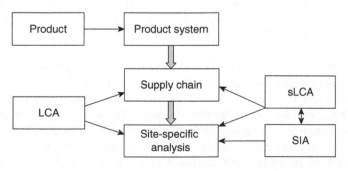

Figure 6.1 Analysis of the product system (Diaz-Chavez, 2012a).

sustainability criteria included in the EC Renewable Energy Directive (2009). Davis et al. (2011) and Diaz–Chavez (2012b) present a review of the different models and the integration of different systems when using land (feed, fodder, fibre, food). Land use planning and zoning have been incorporated in the programmes and policies of different countries when considering a sustainable production of crops for bioenergy use. This is also one of the recommendations of the RSB (see guidelines, mentioned above) and the GBEP. GBEP with the UN have managed to collaborate with the Economic Community Of West African States (ECOWAS) regarding the zoning for energy crops in several countries in sub-Saharan Africa (GBEP, 2012). See also Chapter 7 on land use.

Feedstock production is related to land use and indirect land use change (iLUC). In the UK ARUP (2014) produced a list-based approach for supporting low iLUC feedstock under the RED criteria. This is intended to promote greater harmonization (2009) among Member States. Around 28 feedstocks were assessed and presented in fact sheets. The lists are proposed to count double (or quadruple) towards national renewable transport target and/or count towards a 2020 sub-target for advanced biofuels (ARUP, 2014). In terms of GHG, the majority of routes selected contribute to an 80% reduction. The authors (ARUP, 2014) group the main findings into four categories.

1   *Feedstock availability*: municipal solid waste (MSW) and commercial and industrial wastes, straw, manures, forestry and renewable electricity have the largest supply potentials while wine residues, tall oil pitch and crude glycerine resources are the most limited. Energy crops, short rotation forestry and algae will be limited after 2020.

2   *Technology*: the report and annex present a range of technologies already available at pilot level (lignocellulosic ethanol and butanol, pyrolysis oil upgrading) and others at commercial level but that are not compatible with the feedstocks list (e.g. biomethane from anaerobic digestion of MSW).

3   *Economic*: wastes with a gate fee have a negative price, and energy-dense feedstocks (e.g. tall oil pitch, crude glycerine, used cooking oil and animal fats) have the highest positive prices.

4   *Competition*: competing uses vary widely depending on the feedstock and the use (e.g. for biofuels). The authors suggest that those feedstocks with minimal expansion potential and high competition levels are likely to suffer price increases if diverted to biofuels.

Additionally, from the social point of view, land tenure has been increasingly considered in sustainability assessments. Changes in land use and land ownership have not always been accompanied by appropriate reforms in policies (Kagwanja, 2006). This has been identified as one of the main social problems mainly in sub-Saharan Africa related to biofuel production. The Land Matrix Database (2014) is an independent initiative for land monitoring that promotes transparency and accountability on land and investment. The database provides information about 'deals' done privately or publicly on land acquisition.

The data includes regions, countries, company or government information, area of land acquired and purpose. A review by Hamelinck (2013) of the Land Matrix Database concluded that only 0.5% of the total 38.3 Mha of land deals worldwide are related to biofuel production.

FAO (2012) produced a guideline on the Responsible Governance of Tenure of Land, Fisheries and Forests in the context of the National Food Security program. Some countries adopted the global guidelines on tenure of land, forests and fisheries to safeguard the rights of their people. The guideline addresses among other issues recognition and protection of legitimate tenure rights, even under informal systems, best practices for registration and transfer of tenure rights, managing expropriations and restitution of land to evicted people who were forcibly evicted in the past, rights of indigenous communities, transparency in land acquisition and mechanisms for resolving disputes over tenure rights.

Multi-criteria assessment and indicators have also grown in use for sustainability assessment. The methodologies have been used not just for assessing feedstocks but also for assessing technologies and industrial development and for monitoring impacts produced from the biofuels and bioenergy sector. Diaz-Chavez (2014) presented a review of indicators and their application for monitoring socio-economic impacts of biofuel production. Multi-criteria assessment has been used also for evaluation of the industry. It considers multiple measures and uses different units that can be qualitative and quantitative. The criteria and indicators used can be selected from well-known frameworks (e.g. UN sustainability indicators; UN, 2007) or agreed with the help of experts and decision makers. With the bioeconomy sector getting more economic support in the USA and Europe, several initiatives have been looking at the sustainability assessment of biorefineries. The EU funded three projects (BIOCORE, EUBIOREF and SUPRABIO) where sustainability assessment was considered. Other methodologies can also be found in the literature (e.g. Sacramento-Rivero, 2012; Cherubini et al., 2009) where multi-criteria assessment was used.

Criteria and indicators are also widely used in sustainability frameworks and standards. Their use expanded after the UN Conference in Rio de Janeiro in 2000. The OECD also used indicators to monitor the state of the environment of the countries during the late 1990s and early 2000s. Some of them were also linked to the policies of governments regarding environmental protection and sustainability progress. The following section reviews some of these frameworks and commodity standards.

Legal systems and standards

Policies and programmes regarding the mandatory use of biofuels as a form to reduce GHG emissions and to promote energy security have evolved at global level. There have been several reviews regarding mandatory targets for biofuels. The Global Renewables Fuel Alliance (GRFA, 2014) keeps an updated map with the global mandates. These vary from 2 to 10%, although several countries have lowered their original targets. Other reviews have been produced by region, for example in Africa (Jumbe et al., 2013; Diaz-Chavez et al., 2010), Latin America (Ascher et al., 2009; Solomon and Bailis, 2014) and Asia (Manalilis et al., 2014).

Although the policies and mandatory targets have influenced the production of biofuels, few legal systems incorporate criteria or indicators that regulate a sustainable production. The EC Renewable Energy Directive (2009) introduced basic sustainability criteria for biofuel production, mainly on environmental issues (GHG, land with high biodiversity value, land with high carbon stock and agro-environmental practices), and has since then approved several standards that cover it and additional criteria (e.g. social). There is ample literature on standards and benchmarking (e.g. see van Dam, 2010; Diaz-Chavez, 2011, 2014; Moser et al., 2014). There are currently 19 voluntary sustainability schemes approved by the EU that apply to 27 Member States. Some of these standards were adapted to comply with the EU RED criteria in 2014 (Table 6.2). Additionally, the European Commission (EC) collaborates with the European Committee for Standardization (ECN) to develop standards for the use of biofuels. The International Organization for Standardization (ISO) is also developing a standard on bioenergy.

Table 6.2 Voluntary sustainability schemes approved by the EC for compliance with the RED criteria

No.	Acronym	Name	Commodity
1	ISCC	International Sustainability and Carbon Certification	All feedstocks, all supply chain
2	Bonsucro	BonsucroEU	Sugar cane and derived products
3	RTRS	Roundtable on Responsible Soy EU	Soy-based products
4	RSB	Roundtable on Sustainable Biomaterials EU	Biomass for biofuels and biomaterials
5	2BSvs	Biomass Biofuels voluntary scheme	Wide range of biofuels
6	RBSA	Abengoa RED Bioenergy Sustainability Assurance	All feedstocks for biofuel production
7	Greenergy	Greenergy Brazilian Bioethanol verification programme	Bioethanol from sugar cane from Brazil
8	Ensus	Voluntary scheme under RED for Ensus bioethanol production	Feed wheat for the production of bioethanol
9	Red Tractor	Red Tractor Farm Assurance Combinable Crops & Sugar Beet Scheme	Cereals, oil seeds and sugar beet produced in the UK up to the first point of delivery of these crops

10	SQC	Scottish Quality Farm Assured Combinable Crops (SQC) scheme	Winter wheat, maize and oil seed rape produced in the north of Great Britain up to the first point of delivery of these crops
11	Red Cert	Red Cert	Wide range of different biofuels and bioliquids
12	NTA 8080	NTA 8080 Dutch Scheme	Wide range of different biofuels and bioliquids
13	RSPO RED	Roundtable on Sustainable Palm Oil RED	Wide range of different biofuels and bioliquids
14	Biograce	Biograce GHG calculation tool	To calculate GHG emissions for a wide range of different biofuels and bioliquids
15	HVO	Renewable Diesel Scheme for Verification of Compliance with the RED sustainability criteria for biofuels	All feedstocks suitable for HVO-type biodiesel (including crude palm oil, rapeseed oil, soybean oil and animal fats) with global coverage
16	Gafta	Gafta Trade Assurance Scheme	All feedstocks suitable for biofuel production
17	KZR INIG	KZR INIG System	Raw materials cultivated and harvested in the EU as well as wastes and residues from the EU

Source: EC (2014).

Another scheme on sustainability of biomass driven by government is the Renewable Fuel Standard (RFS) in the USA under the Environment Policy Act of 2005. The revised RFS2, produced in 2010, provides a methodology to assess the life cycle of GHG emissions. The state of California has a separate standard called the Low Carbon Fuel Standard which has its own methodology for GHG assessment and its own carbon reduction targets (Moser et al., 2014).

The majority of voluntary schemes present principles, criteria and indicators on similar issues: environment, social, economic and governance. Table 6.3 includes a summary of selected sustainability schemes applicable to biomass. The inclusion of the GBEP is deemed important due to the comprehensive set of indicators that includes the three main pillars of sustainability (environmental, economic, social) and a fourth topic on energy security (FAO, 2011).

Table 6.3 includes some of the standards and frameworks that are related to solid biomass (FSC, Global Standard, GBEP). The Green Gold Label (GGL) programme is a certification scheme for biomass aimed to producers of agriculture, forest and related industries. It works as a meta-standard because the biomass coming from agriculture or forestry requires to be certified by approved standards (e.g. FSC, PEFRC, GlobalGap). The Forest Stewardship Council (FSC) standard and the Program for Endorsement of Forest Certification (PEFC) are the two

Table 6.3 Summary of selected sustainability standards and frameworks related to biomass on socio-economic and environmental criteria

Standard/ framework	Socio-economic principles	Environmental principles
RTRS	Legal compliance and good business practice Responsible labour conditions Responsible community relations Environmental responsibility Good agricultural practice	Environmental responsibility Good agricultural practice
RSPO	Commitment to transparency Compliance with applicable laws and regulations Commitment to long-term economic and financial viability Use of appropriate best practices by growers and millers Responsible consideration of employees and of individuals and communities affected by growers and mills Responsible development of new plantings Commitment to continuous improvement in key areas of activity	Use of appropriate best practices by growers and millers Environmental responsibility and conservation of natural resources and biodiversity Responsible development of new plantings
FSC	Compliance with laws and FSC principles Tenure and use rights and responsibilities Indigenous peoples' rights Community relations and worker's rights Benefits from the forests: ensure economic viability and a wide range of environmental and social benefits Management plan Monitoring and assessment: to assess activities and social and environmental impacts Maintenance of high conservation value forests Plantations shall be planned and managed	Environmental impact Monitoring and assessment Maintenance of high conservation value forests Plantations
RSB	Planning with impact assessment and management process and an economic viability analysis Not violate human rights or labour rights, and shall promote decent work and the well-being of workers Contribute to the social and economic development of local, rural and indigenous people and communities	Planning, monitoring and continuous improvement Greenhouse gas emissions Conservation Soil Water Air

	Biofuel operations shall ensure the human right to adequate food and improve food security in food insecure regions Maximize production efficiency and social and environmental performance, and minimize the risk of damages to the environment and people Biofuel operations shall respect land rights and land use rights	
Bonsucro	Obey the law Respect human rights and labour standards Manage input, production and processing efficiencies to enhance sustainability Actively manage biodiversity and ecosystem services Continuously improve key areas of the business	Manage input, production and processing efficiencies to enhance sustainability Actively manage biodiversity and ecosystem services Continuously improve key areas of the business
ISCC	Good social practice regarding human rights/labour rights compliance Land rights compliance Priority for food supply/food security	Biomass shall not be produced on land with high biodiversity value or high carbon stock and not from peat land (according to Article 17(3) of the Directive 2009/28/EC and § 4 to 6 of the German BioSt-NachV and BioKraft-NachV). HCV areas shall be protected. Biomass shall be produced in an environmentally responsible way. This includes the protection of soil, water and air and the application of Good Agricultural Practices
GBEP		Greenhouse gas emissions Productive capacity of the land and ecosystems Air Quality Water Availability, use efficiency and quality Biological Diversity Land use change, including indirect effects
GGL Forest	Long-term tenure and use rights to the land and forest resources Management plan	Environmental impact Monitoring and assessment Plantations Other sources than natural forests and plantations (lanes, parks, woods <5 has)

main standards for certification of forestry products. The standards for solid bio-mass are scarce and until now the debate in the EU has continued on minimum sustainability criteria for bioenergy use or advanced biofuel production.

Costs associated with the use of sustainability standards

While the intent of sustainability assurance standards may be to result in positive economic, social and environmental effects across the supply chain, their actual impacts still require analysis. One of the most significant impacts of the implementation of certification schemes on biofuel feedstock production is the added costs of certification. These costs are case-specific and variable depending on the producer, where smallholder producers tend to be more affected (Narayane, 2012). To obtain a certification, the producer must go through the systematic process of the standard which requires examination from an independent auditor of the whole production chain (Europa, 2011). Narayane (2012) compared two sources where the major costs associated with obtaining certification were analysed (Huay Lee et al., 2011 and the BEFSCI, 2012). Both sources agreed that the cost of certification can be broken down into three main types: compliance costs, transaction costs and opportunity costs (see Table 6.4).

Table 6.4 Major costs associated with certification of biofuels at the operator level (after Narayane, 2012)

Type of cost	Huey Lee et al. (2011)	BEFSCI (2012)
Compliance	Associated to the actual implementation of the standard, for example: • change from child to adult labour, • pest management techniques, • training of personnel.	Increased costs due to: • more time and labour required for new production methods, • training activities for producers.
Transaction	Payments during the certification process such as: • third-party inspection fees, • administrative charges for certification process, • time allocated for certification process in smallholders.	Costs associated with the processes and institutions: • achieving certification, • managing the certification process.
Opportunity	Missed opportunities by farmers as consequence of standard implementation: • when farm productivity is affected by sustainable practices such as reduction of fertilizer.	Costs associated with standard implementation: • leaving a portion of land for conservation rather than production.

Sources: Huay Lee et al. (2011), BEFSCI (2012).

Having pointed out some of the implications of implementing voluntary standards, it is not possible to deny that it may also bring benefits. Some of these benefits include: access to markets, improved image, credibility and differentiation from other products (BESCI, 2012). Apart from the costs involved, financial benefits are incurred. At the producer level, there may be increased revenue gains due to:

- reduction of inputs needed, due to improved process efficiencies,
- increase in yields, due to improved production methods,
- increased productivity,
- increased product quality,
- price premiums for certified products (International Trade Centre, 2010).

The extent of the benefits or burdens of the implementation of certification schemes vary according to context. The major factors which make the costs tolerable for producers are economies of scale and the preparedness of exporters. Nevertheless, it is important to differentiate between implementing a standard, verifying and auditing it to obtain a certification (Woods and Diaz-Chavez, 2007). Figure 6.2 shows the main steps for the effective flowing of a certification scheme.

Although the EU RED Directive has encouraged the biofuel market and made it attractive for developing countries, large-scale producers can more easily bear the costs involved with certification, the logistical requirements and

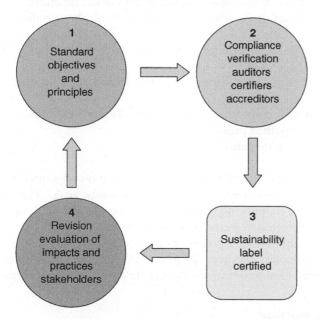

Figure 6.2 Process flow for the effective functioning of a certification scheme (after Steering Committee of the State-of-Knowledge Assessment of Standards and Certification, 2012).

adapt to necessary changes in production practices. On the other hand, small-holder farmers may face high transaction costs and a lack of capital, buyers, technology or information, which makes certification very difficult.

For the integration of these small farmers into such certification schemes, financial and social incentives are necessary (Huay Lee et al., 2011). FAO stated that costs and complexities associated with certification may inhibit the incorporation of small producers into the bioenergy value chain. There are currently very few schemes that include specific criteria to encourage the inclusion of small farmers (e.g. RSB, Fairtrade) (BEFSCI, 2012). The International Trade Centre (2010) has indicated that certification requirements may encourage changes in organizational structures and production processes. This allows for improvement of skills such as management and monitoring systems, good resource management, etc., which can yield to other benefits such as better access to credit. This may encourage more skilled labour, but the impact on the numbers employed is still unknown. Improved processes may also imply improved working conditions for employees (Narayane, 2012).

Narayane (2012) investigated the barriers of certification in Brazil sugar cane and soy production for biofuels and found other associated costs and barriers to certification at the production level (industrial), including:

- need for staff dedicated to sustainable production and to oversee the certification process within the organizations,
- lack of capacity within these companies in the field of sustainability,
- time taken depends on the availability of auditors with the specific time required,
- surveillance audits and subsequent compliance audits to maintain the certification,
- barriers associated with compliance with the Brazilian environmental legislation,
- barriers associated with compliance with Brazilian labour legislation,
- lack of demand for certified products,
- lack of incentives for certification.

The barriers and benefits of the implementation of certification schemes do not necessarily explain the achievement of broader sustainability goals. This will require a different type of monitoring that is not related to the use of voluntary standards. It is related to the achievement of other broader sustainability goals and the monitoring of impacts either positive or negative. At the same time, this is also related to the voluntary reporting of private companies. The following section reviews the need of monitoring and the benefits of sustainable reporting.

Monitoring and reporting

Most of the voluntary standards reviewed work on a compliance basis but in terms of sustainability few of these indicators can be monitored over time and under quantitative or clear qualitative parameters. The main aim of sustainability indicators is to simplify information; they can be useful to show change patterns on economy, the environment and society on a baseline (Diaz-Chavez, 2014). Agenda 21 (UNCSD, 2012) indicates that the function of indicators is to provide a basis for planning and decision making. As in voluntary standards and sustainability schemes, indicators can contribute to monitor impacts (both positive and negative) of certain activities or over a period of time. This was the case of the OECD environmental performance indicators at country level (OECD, 2004).

The Global-Bio-Pact EU FP7-funded project developed a set of indicators to monitor socio-economic impacts of biofuel production that could be tested on the field (Diaz-Chavez and Vuohelainen, 2014). The set of indicators was applied in two case studies in Brazil and Argentina. The set of indicators included methodologies to obtain data from the government (e.g. census), from the companies and plantations (e.g. workers, investors, owners), and from the local communities and NGOs. The use of surveys probed to be a valuable tool to gather information especially from the perceptions of the communities. Similar indicators have been used in Brazil to assess the sugar cane industry and whether social indicators such as education, employment (quality and quantity) and wages have improved (Moraes, 2011).

Furthermore, monitoring over time needs to be done at local or regional level to be able to observe changes in sustainable production. For the field test of the Global-Bio-Pact indicators, producers were requested to provide information from 5 years prior to the assessment, but this information had either not been collected or was not easily accessible. The gathering of data in the field was done through lists similar to those used in an environmental audit or verification system. Most of the indicators to measure impacts can be measured quantitatively, but some of the indicators are qualitative and thus somewhat more difficult to measure.

The assessment of the indicators showed that most of the indicators were clear and easily understandable for the respondents. The operators agreed that keeping records of the information would be useful for monitoring the impacts of the operation. The field test demonstrated that operators have different ways of capturing data, which can make it difficult for getting standardized information across different operations even in the same area (Diaz-Chavez and Vuohelainen, 2014). The issue of availability of data would probably be solved if the indicators were applied in a more formalized way, e.g. as a part of certification scheme, and the operations would have systems in place to routinely collect the information from their operations. The national level can be used for background information. For monitoring impacts, it is necessary to set clear sustainability goals and a baseline. It should also be considered that bioenergy involves agricultural and agroindustrial activities in a region and it is very difficult to differentiate the impact of the biofuels or bioenergy production from those of the whole system. This is especially challenging for mixed food/fuel crops as, considered in this study, both Diaz-Chavez and Vuohelainen (2014) concluded that a system engaging the government, the private sector, NGOs and the communities would be the way forward for monitoring. Nevertheless, this would require human and economic resources that would be difficult to cover by the public and private sectors.

Another form of monitoring that continues to improve is voluntary reporting. Some private companies have included sustainability reporting in their Corporate Social Responsibility goals. The main organization and framework is from the Global Reporting Initiative (GRI). GRI promotes the use of sustainability reporting as a way for organizations to become more sustainable and contribute to sustainable development (GRI, 2014). The GRI provides Sustainability Reporting Guidelines for the preparation of sustainability reports by organizations. The guidelines offer an international reference for all those interested in the disclosure of governance approach and of the environmental, social and economic performance and impacts of their organizations.

The guidelines are developed through a global multi-stakeholder process involving different stakeholders from business, labour, civil society and financial markets. The process also involves auditors and experts in various fields, and regulators and governmental agencies in several countries. The GRI framework and Sustainability Reporting Guidelines set out the principles and standard disclosures that organizations can use to report their economic, environmental and social performance and impacts (GRI, 2014). The framework refers to materiality as the information contained in a report that covers topics and indicators which reflect the organization's impacts. The categories include aspects, which are the list of subjects covered by the guidelines. The economic category, for instance, includes economic performance, market presence, indirect economic impacts and procurement among its aspects. The environmental category includes aspects such as energy, water, biodiversity, emissions and waste, among others. The social category includes labour practices and decent work (e.g. employment), human rights (e.g. non-discrimination, freedom of association), society (e.g. local communities, anti-corruption) and product responsibility (e.g. customer health and safety, product service and labelling) (GRI, 2013).

The GRI framework is also applicable by sector. This is both a challenge and a bias since the indicators will apply in a different form. Some companies with activities in energy, agriculture and forestry have started reporting under the GRI. There are currently 6,272 organizations, 18,529 reports and 15,013 GRI reports registered in GRI. For instance, sugar cane mills in Colombia and Brazil and soy mills in Argentina have published their reports in the GRI website (GRI, 2014).

Although sustainability reporting is not yet common in the private sector, it might be a possible way forward to monitor the activities of the private sector. Nevertheless, it would benefit the reporting system if a minimum number of indicators for the bioenergy system is covered.

Conclusions

There is not one single methodology that would cover the synergies of environmental, economic, social and governance issues required to assess the sustainable production and use of bioenergy systems. Most focus of sustainability assessment in this sector has been given to the feedstock production, particularly for biofuels production. This chapter reviewed some of the most common methodologies and tools used for sustainability assessment although it is not exhaustive. Methodologies continue to derive from available ones but in reality they are mainly the same ones with some modifications. Appendix 6.1 provides further details.

Life-cycle analysis and environmental management tools are a form to combine data for a sustainability assessment. This continues to be a more common form to assess projects that cover the whole supply chain. Particularly, when the production system is involved. Both methodologies use indicators although the application and type of data may be a challenge for a combined methodology. This is particularly for those indicators that are qualitative in nature.

Multi-criteria assessment is deemed as one of the methodologies that provide a whole picture for an integrated assessment. It also faces the problem of having a majority of environmental indicators which are quantitative in nature and thus easily measured while socio-economic indicators tend to have more qualitative indicators that make the assessment more difficult. Nevertheless, both should still be considered in the assessment. As demonstrated with practical projects, some qualitative indicators could be further standardized in terms of the information requested making them easier to measure and compare across time scales and even across regions or feedstocks.

Operators can be engaged in the activities of monitoring if they receive advice on how this information could then collected annually so as to monitor changes in the indicators. It could be possible to ask operators to report annually on a subset of the indicators. This would make it possible to create a monitoring system for small operators and could be combined with efforts done with local governments.

Monitoring on the basis of Corporate Social Responsibility is gaining attention but it is still kept within large companies. Incentives from the public sector would contribute to improving the conditions for reporting and making publicly available the information without the companies' activities and sensible information being compromised.

References and further reading

ARUP (2014) *Advanced biofuel feedstocks: an assessment of sustainability.* ARUP Consortium and E4Tech. Framework for Transport-Related Technical and Engineering Advice and Research (PPRO 04/45/12). Department for Transport, London.

ASCHER, M., GANDUGLIA, F., VEGA, O., ABREU, F. AND MACEDO, J. (2009) *Study on regional evidence generation and policy and institutional mapping on food and bioenergy: Latin America and Caribbean-LAC.* Paper prepared on behalf of FORAGRO/

PROCITROPICOS/IICA. Presented at the ERA-ARD/SAG Special Event on Bioenergy Brussels, December 2009.

BEALL, E., CADONI, P. AND ROSSI, A. (2012) *A compilation of tools and methodologies to assess the sustainability of modern bioenergy.* Food and Agriculture Organization. Environment and Natural Resources working paper no.51 2012. FAO, Rome.

BEFSCI (2012) *Smallholders in Global Bioenergy Value Chains and Certification Bioenergy and food security criteria and indicators project.* Food and Agriculture Organization of the United Nations, Environment and natural resources management working paper

BLACK, M., WHITTAKER. B.C., HOSSEINID, S.A., DIAZ-CHAVEZ, R., WOODS, J. AND MURPHY, R.J. (2011) Life Cycle Assessment and sustainability methodologies for assessing industrial crops, processes and end products. *Industrial Crops and Products* 34, 1332–1339.

CHERUBINI, F., JUNGMEIER, G., WELLISCH, M., WILLKE, T., SKIADAS, I., VAN REE, R. ET AL. (2009) Towards a common classification approach for biorefinery systems. *Biofuels, Bioproducts and Biorefining* 3, 534–546.

DAVIS, C.S., HOUSE, J.I., DIAZ-CHAVEZ, R., MOLNAR, A., VALIN, H. AND DELUCIA, E. (2011) How can land-use modelling tools inform bioenergy policies? *Interface Focus* 1, 212–223.

DIAZ-CHAVEZ, R. (2008) *Updated sustainability assessment: a comparison of BEST sites 2007–2008.* BEST Deliverable 9. 23 January 2008, London. Bioethanol for Sustainable Transport FP6 EC project. TREN/05/FP6EN/S07.53807/019854.

DIAZ-CHAVEZ, R. (2011) Assessing biofuels: aiming for sustainable development or complying with the market? *Energy Policy* 39, 5763–5769.

DIAZ-CHAVEZ, R. (2012a) *WP7 sustainability assessment and socio-economic methodology.* BIOCORE EU FP7 project.

DIAZ-CHAVEZ, R. (2012b) Land use for integrated systems: a bioenergy perspective. *Environmental Development* 3, 91–99.

DIAZ-CHAVEZ, R. (2013) *Socio-economic assessment of biorefineries concept.* BIOCORE EU FP7 project.

DIAZ-CHAVEZ, R. (2014) Indicators for socio-economic sustainability assessment. In Janssen, R. and Rutz, D. (eds), *Socio-economic impacts of bioenergy production.* Springer, Berlin, chapter 2.

DIAZ-CHAVEZ, R. AND VUOHELAINEN, A. (2014) Test auditing of socio–economic indicators for biofuel production. In Janssen, R. and Rutz, D. (eds), *Socio-economic impacts of bioenergy production.* Springer, Berlin, chapter 3.

DIAZ-CHAVEZ, R., MUTIMBA, S., WATSON, H., RODRIGUEZ-SANCHEZ, S. AND NGUER, M. (2010) *Mapping food and bioenergy in Africa: a report prepared for FARA.* Accra, Ghana.

EC (2009) Renewable Energy *Directive 2009/28/EC of the European parliament and of the Council of 23 April 2009,* Official Journal of the European Union.

EC (2010) *International Reference Life Cycle Data System (ILCD) handbook: general guide for life cycle assessment: detailed guidance.* Joint Research Centre, Institute for Environment and Sustainability. EUR 24708 EN. Publications Office of the European Union, Luxembourg.

EC (2014) *Renewable energy biofuels: sustainability schemes.* http://ec.europa.eu/energy/renewables/biofuels/sustainability_schemes_en.htm.

ENGLUND, O., BERNDES, G., JOHNSON, H. AND OSTWALD, M. (2011) *Environmental impact assessment: suitable for supporting assessments of biofuel sustainability?* Technical report for the EU Biofuel Baseline project. Chalmers University of Technology, Gothenburg. http://publications.lib.chalmers.se/records/fulltext/local_146738.pdf.

EUROPA (2011) *Commission sets up system for certifying sustainable biofuels.* http://europa.eu/rapid/pressReleasesAction.do?reference=MEMO/11/522&format=HTML.

FAO (2011) GBEP *The Global Bioenergy Partnership sustainability indicators for bioenergy*, 1st ed. Food and Agriculture Organisation, Rome.

FAO (2012) *FAO voluntary guidelines on the responsible governance of tenure of land, fisheries and forests in the context of national food security.* Food and Agriculture Organisation. www.fao.org/docrep/016/i2801e/i2801e.pdf.

GBEP (2010) *Analytical tools to assess and unlock sustainable bioenergy potential.* The Global Bioenergy Partnership. FAO, Rome.

GBEP (2012) *ECOWAS Regional Bioenergy Forum.* Bamako, Mali 19–20 March 2012. www.globalbioenergy.org/fileadmin/user_upload/gbep/docs/2012_events/WGCB_Activity_1_Bamako_19-22_March_2012/ECOWAS_Regional_Bioenergy_Forum_Concept_Note.pdf.

GIBSON, R. (2005) *Sustainability assessment.* Earthscan, London.

GRFA (2014) *Global Renewables Fuel Alliance.* http://globalrfa.org/news-media/.

GRI (2013) *G4 sustainability reporting guidelines.* Global Reporting Initiative. https://www.globalreporting.org/resourcelibrary/GRIG4-Part1-Reporting-Principles-and-Standard-Disclosures.pdf.

GRI (2014) *Sustainability reporting database.* Global Reporting Initiative. http://database.globalreporting.org/search.

HAMELINCK, C. (2013) Land grabs for biofuels driven by EU biofuels policies. Ecofys. www.ecofys.com/files/files/ecofys-2013-report-on-land-grabbing-for-biofuels.pdf.

HUAY LEE, J.S., RIST, L., OBIDZINSKI, K., GHAZOUL, J. AND PIN KOH, L. (2011) No farmer left behind in sustainable biofuel production. *Biological Conservation* 144, 2512–2516.

INTERNATIONAL TRADE CENTRE (2010) *Market access, transparency and fairness in global trade export impact for good.* Geneva. http://legacy.intracen.org/publications/Free-publications/Market-access-transparency-fairness-in-global-trade-Export-Impact-for-Good-2010.pdf.

JUMBE, C. AND MKONDIWA, M. (2013) Comparative analysis of biofuels policy development in Sub-Saharan Africa: the place of private and public sectors. *Renewable Energy* 50, 614–620.

KAGWANJA, J. (2006) Land tenure, land reform, and the management of land and natural resources in Africa. In Mwangi, E. (ed.), *Land rights for African development: from knowledge to action.* Collective Action and Property Rights (CAPRi), Washington DC. http://knowledgebase.terrafrica.org/fileadmin/user_upload/terrafrica/docs/brief_land.pdf#page=5.

LAND MATRIX DATABASE (2014) *The online public database on land deals.* http://landmatrix.org/.

MANALILIS, N., BADAYOS, R. AND GOUR, V.K. (2014) *Regional evidence generation and policy and institutional mapping on food and bioenergy (Asia).* Asia-Pacific Association of Agricultural Research Institutions (APAARI). www.era-ard.org/fileadmin/SITE_MASTER/content/Dokumente/APAARI_presentation.pdf.

MORAES, M.A.F.D. (2011) Socio-economic indicators and determinants of the income of workers in sugar cane plantations and in the sugar and ethanol industries in the north, north-east and centre-south regions of Brazil. In Amann, E., Baer, W. and Coes, D. (eds), *Energy, bio fuels and development: comparing Brazil and the United States.* Routledge, Taylor and Francis Group, pp. 137–150.

MOSER, C., HILDEBRANDT, T. AND BAILIS, R. (2014) International sustainability standards and certification. In Solomon, B. and Bailis, R. (eds), *Sustainable development of biofuels in Latin America and the Caribbean.* Springer, Berlin, pp. 27–70.

NARAYANE, R.M. (2012) *Socio-economic implications of sustainability assurance standards for biofuels produced in developing countries for use in the EU: the case study of Brazil.* Thesis for the Master of Science in Sustainable Energy Futures, Energy Futures Lab., Imperial College London.

NATURAL ENGLAND (2014) *The Environmental Impact Assessment (Agriculture) (England) (No.2) Regulations 2006 Act.* www.naturalengland.org.uk/ourwork/regulation/eia/default.aspx.

OECD (2004) *Key environmental indicators.* Organization for Economic Co-operation and Development, Paris.

OECD (2011) *Strategic environmental assessment and biofuel development.* SEA Toolkit. Organization for Economic Co-operation and Development, Paris.

RSB (2011) *RSB impact assessment guidelines. Roundtable on sustainable biopfuels.* RSB reference code RSB-GUI-01-002-01 (version 2.0). rsb.org/pdfs/guidelines/11-03-09%20RSB-GUI-01-002-01%28RSB-IA-Guidelines%29.pdf.

SACRAMENTO-RIVERO, J. (2012) A methodology for evaluating the sustainability of biorefineries: framework and indicators. *Biofuels, Bioproducts and Biorefining* 6, 32–44.

SHEATE, W. (ed) (2012) *Tools, techniques and approaches for sustainability. Collected writings in environmental assessment policy and management.* World Scientific, pp. 1–32.

SHD (SOCIAL HOTSPOTS DATABASE) (2015) *Social hotspots database.* http://social-hotspot.org/.

SOLOMON, B. AND BAILIS, R. (eds) (2014) *Sustainable development of biofuels in Latin America and the Caribbean.* Springer, Berlin.

STEERING COMMITTEE OF THE STATE-OF-KNOWLEDGE ASSESSMENT OF STANDARDS AND CERTIFICATION (2012) *Toward sustainability: the roles and limitations of certification.* Executive summary. RESOLVE, Washington DC.

UN (2007) *Indicators of sustainable development: guidelines and methodologies October 2007,* 3rd edn. www.un.org/esa/sustdev/natlinfo/indicators/guidelines.pdf.

UNCSD (2012) *Sustainable development in the 21st century (SD21).* Review of implementation of Agenda 21 and the Rio Principles Synthesis, January 2012. Study prepared by Stakeholder Forum for a Sustainable Future. www.uncsd2012.org/content/documents/194Synthesis%20Agenda%2021%20and%20Rio%20principles.pdf.

UNEP (2009) *Guidelines for social life cycle assessment of products,* Benoît, C. and Mazijn, B. (eds). Belgium.

VALDIVIA, S., UGAYA, C., HILDENBRAND, J., TRAVERSO, M., MAZIJN, B. AND SONNEMANN, G. (2013) A UNEP/SETAC approach towards a life cycle sustainability assessment: our contribution to Rio+20. *International Journal of Life Cycle Assessment* 18(9), 1673–1685.

VAN DAM, J. (2010), *Background document.* From van Dam, J., Junginger, M. and Faaij, A. (2010) From the global efforts on certification of bioenergy towards an integrated approach based on sustainable land use planning. *Renewable and Sustainable Energy Reviews* 14(9), 2445–2472.

WECD (1987) *Report of the world commission on environment and development: our common future.* United Nations. http://conspect.nl/pdf/Our_Common_Future-Brundtland_Report_1987.pdf.

WOLF, M.A., PANT, R., CHOMKHAMSRI, K., SALA, S. AND PENNINGTON, D. (2012) *Life Cycle Data System (ILCD) handbook.* JRC Reference Reports. Joint Research Centre, Italy.

WOODS, J. AND DIAZ-CHAVEZ, R. (2007) *The environmental certification of biofuels.* Report for the OECD. OECD, Paris.

WSDD (2002) *Report of the world summit on sustainable development.* http://daccess-dds-ny.un.org/doc/UNDOC/GEN/N02/636/93/PDF/N0263693.pdf?OpenElement.

Appendix 6.1   Methodology for sustainability assessment criteria and indicators selection for bioenergy

Different methods and frameworks can be used to assess the sustainability for a bioenergy project. There is no single best methodology for conducting a sustainability appraisal or assessment. Rather, this requires the use of a wide range of analytical tools which derive from EIA, policy analysis and plan evaluation practice, among others. As suggested by the UN (2007), the type of analytical tools to be used depends on the type of sustainability assessment required. For instance, EIA and SIA methods can be adapted where a cause–effect chain can be identified (i.e. checklists, GIS, predictive modelling, LCA, impact networks, multi-criteria assessment). This appendix presents two examples of the selection of indicators and assessment.

The International Association for Impact Assessment defines an impact assessment as 'a structured . . . process for considering the implications, for people and their environment, of proposed actions while there is still an opportunity to modify (or even, if appropriate, abandon) the proposals. It is applied at all levels of decision making, from policies to specific projects.'[1] There are several types of impact assessment, which cover a wide range of issues, including environmental or social impact assessments. A sustainability assessment covers economic, social and environmental issues, and also policy and regulations or governance issues. It looks at the synergies among them and can establish a cause–effect link but most importantly will cover aspects of stakeholder identification and participation.

Having set the objectives of the assessment (e.g. assessment of a supply chain, assessment of a project or a region) a matrix can be used to first select those indicators considered to be the most representative. Table 6.5 shows some of these considerations.

One condition that must always be considered is that the ideal indicator does not exist. When selecting data for indicators, if data is not available, a second-best proxy is often used to replace the original indicator (Segnestam, 1999). Diaz-Chavez (2011) developed a matrix for the rapid assessment of a biomass supply change. Table 6.6 shows this matrix as an example for a supply chain of biofuels. Nevertheless, the matrix can be adapted for other supply chains.

In the first step the matrix like that in Table 6.6 can be used to simply select criteria that are important for the supply chain, then allowing the selection of indicators. There are also different forms for selecting the indicators. In the Global–Bio–Pact project (see Diaz–Chavez, 2014b) the following steps were followed for the selection of indicators:

- benchmarking of standards for environmental and social indicators,
- identification of impacts of biomass use for bioenergy/biofuel production previously identified in the project through case studies and background information,

Table 6.5 Criteria and issues to address in a sustainability assessment (modified from Diaz-Chavez, 2011)

Environmental	Social	Economic	Policy and institutional (governance)
Land use	Social organization	Economic value of	National, regional and
Biodiversity	and demographic	resources	local legislation
Resources	characteristics	Local economy	National, regional and
(biomass)	Stakeholder	Production level	local PPPs
Carbon	participation	(small/large scale,	International
conservation	Labourers' rights	family/small owners)	considerations
Soil	Child labour	Investment (funds)	Institutional capacity
Water	Gender (women	Trade (incentives and	Political incentives
(availability	participation)	barriers)	and barriers
and quality)	Land use rights	Market	Lobbying
GHG	Working conditions	Supply chain values	Governance
Air emissions	Livelihoods (non-	Costs of production	Stakeholder mapping
Spatial and	economic value of	Costs of certification	
timeframe	biomass resources)	Scale production	
considerations	Use of resources (FvF)	considerations	
Waste	Health	Climate change risks	
(agriculture	Quality of life	Poverty reduction	
competition)	Rural development	By-products,	
	Education and skills	co-products	
	Technology	Rural development	
	acquisition and transfer		
	Climate change risks,		
	vulnerability and		
	adaptation		

FvF, fuel versus food; PPP, policies, plan programmes.

Table 6.6 Rapid sustainability assessment for biofuel (RaSAB) production

Criteria per chain	Crop growing	Transport	Ethanol plant	Distribution (includes transport to)
GHG				
Feedstock used				
Yield (t/ha)				
Fertilizer (kg/year)				
Pesticide (kg/ha)				
Seeds (kg/ha)				
Energy use drying				

Transport mean and mileage

Environmental

Land use last 5 years

Crop used last 5 years

Irrigation (type and amount of water used)

Agricultural waste (kg/year)

Compliance CAP on GAP

Farm certification on any product

Social

No. of workers

Gender of workers

Ethnicity of workers

Education of workers

Workers per activity

Minimum EU salary or local salary

No. of hours worked/day

Economic

Cost of employees (salaries paid)

Cost of raw material per activity

Cost of electricity/fuel monthly

Cost of current certification (if any)

Source: Diaz-Chavez (2008).

- identification of socio-economic impacts in supply chains,
- links between environmental and social impacts,
- macro and micro indicators in the case studies.

See also Figure 6.3.

Selected indicators

According to the methodology followed and the benchmarking review, it is expected that the set of indicators will be used by different stakeholders:

- in initiating or assessing a bioenergy proposal or project,
- in assessing the sustainability of a feasibility report for a bioenergy proposal or project,

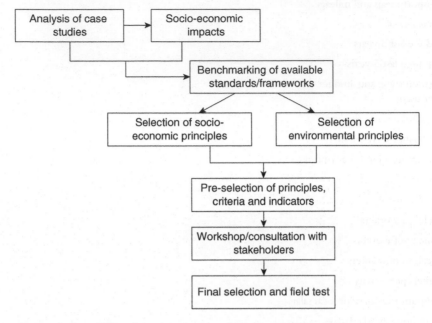

Figure 6.3 Steps followed for the selection of criteria and indicators.

- in monitoring impacts at the local and regional level,
- in addition to a standard.

Finally, the set of indicators may differ under different frameworks, projects, experts, countries or any other stakeholders' opinion. In the case of the Global-Bio-Pact project, the indicators were selected to measure an impact over a period of time. For this reason a baseline was suggested for the field test work.

The indicators were classified in different categories including background information, socio-economic indicators and environmental indicators:

- *basic information*: data that provides background information from the selected case study,
- *socio-economic indicators*: these include the impacts caused by bioenergy crops production and the different stages of the supply chain to produce biofuels,
- *environmental indicators*: in the project's context they refer to the environmental impacts that affect the socio-economic characteristics of the communities.

Table 6.7 presents the main topics and impacts selected. The guideline explanation of the impact is included in the indicators list. For a detailed list of indicators see Diaz–Chavez (2014b).

Table 6.7 Impacts and examples of indicators (Diaz–Chavez, 2014b)

Impact	Examples of indicators
Basic information	
Framework conditions	Location, average yield
Socio-economic	
Contribution to local economy	Value added, employment
Working conditions and rights	Employment benefits
Health and safety	Work-related accidents
Gender	Benefits
Land rights	Land rights and conflicts
Food security	Land converted from staple crops
Environmental	
Air	Open burning
Soil	Soil erosion
Water	Availability of water
Biodiversity	Conservation measures
Ecosystem services	Access to ecosystem services

Methodology for mapping stakeholders

The methodology for mapping stakeholders firstly identifies them at national level. Then stakeholders are identified at the productive level including NGOs, farmers, other civil organizations and the industry sector (including farmers with different forms of participation, e.g. out-growers). A quadrat is considered to include stakeholders from the local government, the national government, NGOs (including other civil organizations) and industry. These last two may also include farmers but at different levels of organization. The links between these different bodies and stakeholders are expressed using lines as direct, indirect or needed and the closer they are the closer the relationship is or should be.

Surveys and interviews can be applied to the identified stakeholders to gather data for the indicators and to include a narrative assessment (Diaz–Chavez, 2014b).

Assessment, uncertainty and subjectivity

Sustainability assessment should seek to minimize the uncertainty in results, in order to provide clearer support for decision making. This can be reduced with the use of different tools and different criteria and indicators. For instance, impact assessment methods in E-LCA attempt to delay value-based aggregation and to keep it separate from what is considered science-based characterization within an impact category, and possible science-based aggregation of impacts within (broader, more encompassing) damage categories. Data that tends to be qualitative can reduce subjectivity, including workers' reports of their perceived degree of control over their schedules and working environment.

Mitigation

If adverse impacts are assessed then mitigation measures will need to be proposed. They can be technical, related to policies and regulations, social or engage the community, among others. For these reasons the SIA and sLCA should consider local participation.

Monitoring

This is probably the most difficult of the different steps to follow as monitoring needs to be incorporated with the Corporate Social Responsibility of the company or developer and linked with the local authorities and local communities. Most of the time this fails due to lack of staff and resources. In the case of BIOCORE a list of measures will be listed and the monitoring programme, including suggestions for applying it, will be developed.

For the analysis, a list of criteria that were used for the case studies in Europe and India is shown in Table 6.8. The quality of life parameter is not considered as it would need reference data and a project that demonstrates actual improvement or a worsening of the situation to be able to conduct any assessment.

In addition, an assessment of hotspots developed by the Social Hotspots Database (SHD)[2] was included according to the corresponding or similar sectors for a biorefinery included in the SHD (e.g. agriculture, forestry, chemical industry, electricity).

The criteria in the BIOCORE indicators were assessed as possible negative and positive impacts and if negative how can they be mitigated or which precautions need to be considered. This assessment does not include a complete social assessment, which requires a well-documented project proposal and would need to comply with national regulations. Nevertheless, it should be noted that a proper ESIA is recommended to comply with good practice (Diaz-Chavez, 2014a).

Testing indicators in the field

A sustainability assessment especially focusing on monitoring was conducted in the Global-Bio-Pact project. While the indicators were tested in the field, four characteristics were selected to assess the effectiveness of the indicators. These were:

- *measurability*: how easy is to measure the impact,
- *ease of gathering data*: how easy and cost-effective it is to gather the data for the indicator,

Table 6.8 Criteria considered for the social sustainability assessment of a biorefinery (Diaz–Chavez, 2014a)

No.	Parameter	Characteristics/criteria	Level of assessment	Stage of the supply chain	Type of data
1	Trade of feedstock	Incentives and barriers	EU/national	Feedstock	Qualitative
2	Identification of stakeholders along the supply chain	Producers (farmers) Regulators Business Traders	Local	All	Qualitative, literature and interviews
3	Policies and regulations	International National Regional Local	National and international	All	Qualitative (section 5)
4	Potential biorefinery location (logistics)	Availability of feedstock Current use of residues Potential for supply of feedstock	National and local	Feedstock, transport and storage, biorefinery	
5	Land use tenure	Land ownership rights	National	Feedstock	Qualitative, literature
6	Community participation	Acceptance of community to: • residues of feedstock used for biorefinery • dedicated feedstock for biorefinery use • biorefinery construction Indigenous communities	Local	Feedstock, transport and storage, biorefinery	Qualitative, interviews Hotspots Database
7	Quality of life	Improvement of quality of life Improvement of livelihoods Improvement of socio–economic conditions	N/A	N/A	N/A

(continued)

Table 6.8 (continued)

No.	Parameter	Characteristics/criteria	Level of assessment	Stage of the supply chain	Type of data
8	Rural development and Infrastructure	Sanitation Roads Water	National level	Feedstock, transport and storage, biorefinery	Qualitative, Hotspots Database
9	Jobs creation and wages	Labour involved on residues collection or dedicated crops Jobs created for biorefinery Jobs created for transportation Wages paid according to national/regional regulation (minimum wage) Poverty reduction	National Local National	Feedstock, transport and storage, biorefinery	Qualitative interviews Hotspots Database, wages Interviews Hotspots Database
10	Gender equity	Inclusion of women	National	Feedstock, transport and storage, biorefinery	Hotspots Database
11	Labour conditions	International Labour Organization conventions and human rights including: • child labour • right to organize • indigenous rights • forced labour	National	Feedstock, transport and storage, biorefinery	Hotspots Database
12	Health and safety	Compliance with health and safety regulations at the different supply chains	National	Feedstock, transport and storage, biorefinery	Qualitative interviews
13	Competition with other sectors	Competition of residues use for biorefinery and impact on other industries and sectors that affects negatively	Local	Feedstock, final products	Qualitative interviews

Source: Modified from BIOCORE D7.4 (Diaz-Chavez, 2014a).

- *usefulness for assessing socio-economic impacts*: whether the indicators really represent the assessment of the impact,
- *temporality*: the time frame for the usefulness of the indicator, including the data (Diaz-Chavez and Vuohelainen, 2014).

Notes

1 Definition by International Association for Impact Assessment: http://www.iaia.org/iaiawiki/impactassessment.ashx.
2 SHD (SOCIAL HOTSPOTS DATABASE) (2015) *Social hotspots database*. http://socialhotspot.org/.

References and further reading

DIAZ-CHAVEZ, R. (2008) Updated sustainability assessment: a comparison of BEST sites 2007–2008. BEST Deliverable 9.23. January 2008, London. Bioethanol for Sustainable Transport FP6 EC project. TREN/05/FP6EN/S07.53807/019854.
DIAZ-CHAVEZ, R. (2011) Assessing biofuels: aiming for sustainable development or complying with the market? *Energy Policy* 39, 5763–5769.
DIAZ-CHAVEZ, R. (2012) Land use for integrated systems: a bioenergy perspective. *Environmental Development* 3, 91–99.
DIAZ-CHAVEZ, R. (2014a) WP7 D7.4 *sustainability assessment and socio-economic assessment*. BIOCORE EU FP7 project. Report.
DIAZ-CHAVEZ, R. (2014b) Indicators for socio-economic sustainability assessment. In Janssen, R. and Rutz, D. (eds). *Socio-economic impacts of bioenergy production*. Springer, Berlin, chapter 2.
DIAZ-CHAVEZ, R. AND VUOHELAINEN, A. (2014) Test auditing of socio-economic indicators for biofuel production. In Janssen, R. and Rutz, D. (eds) *Socio-economic impacts of bioenergy production*. Springer, Berlin, chapter 3.
UN (2007) *Indicators of sustainable development: guidelines and methodologies October 2007*, 3rd edn. www.un.org/esa/sustdev/natlinfo/indicators/guidelines.pdf.

7 Land use assessment for sustainable biomass

Alexandre B. Strapasson, Lei Wang and Nicole Kalas

Introduction

Bioenergy is an opportunity for tackling global warming and promoting sustainable development globally. More than a renewable energy source, bioenergy can be used as a vector for rural development, contributing to a more equitable geopolitical relationship worldwide. However, the confluence of energy and food demands and the increasing scarcity of natural resources impose an increasing need to find new strategies for sustainable bioenergy. The challenge is how to promote bioenergy in a symbiotic way with food production and conservation of ecosystems and their services. Therefore, land use assessments for biomass are fundamental for governing the equitable use of land and the allocation of that land to different productive and non-productive uses.

Energy and food securities are strategic issues for any country and they often supersede options to develop economically and environmentally sustainable bioenergy, which in turn entail extra effort requiring public policies encompassing global responsibilities. World population, for example, is likely to increase from about 7 billion in 2014 to 9.6 billion inhabitants by 2050, in a range between 8.3 and 10.9 billion (UN, 2011). Associated with this United Nations forecast, income per capita is likely to keep increasing in the coming decades, especially in emerging countries with high population, which consequently would lead to a greater demand for food and energy per capita. Rural exodus is also a challenge for cities in developing countries, mainly in Africa and parts of Asia and Latin America. This migration process may intensify urban problems even more (e.g. slums, sewage and waste treatment, public transport, water supply, etc.) due to the lack of infrastructure and job opportunities for people excluded from their rural areas. Therefore, in order to ameliorate these problems, bioenergy must be understood in a broader context, involving energy, agriculture, environmental, social and political perspectives.

Thus, before starting any bioenergy programme, it is important to assess the land use potential for sustainable biomass of a certain region or nation, which would then serve as a basis for the supporting public policies and future bioenergy investments. There are several ways for doing so, e.g. using models or zoning schemes, as presented in the following sections of this chapter.

Models

Increased demand for crop-derived biofuels (bioethanol and biodiesel), in addition to food and feed demand, induces direct and indirect land use changes (LUC) as well as the consequential greenhouse gas (GHG) arising from such changes through various processes such as deforestation, replanting and land ploughing, resulting in loss/gain of carbon stocks and soil organic matter. In the context of biofuels, LUCs are of two types (see also Figure 7.1).

1 Direct land use change (dLUC) occurs when crops for biofuel production are planted on land that has not previously been used for that purpose; for example, the conversion of forest into an energy crop plantation for biofuels production. The effects of dLUC can be directly observed and measured as the effects are localized to a specific plantation.

2 Indirect land use change (iLUC) is considered to occur when, as a result of switching agricultural land use to biofuel crops, a compensating LUC occurs elsewhere to maintain the previous level of agricultural production. These effects are typically the unintended consequence of land use decisions elsewhere and, given that the effects are not limited by geographical boundaries (e.g. the complex dynamics of food commodities worldwide), often are not directly observable or measurable.

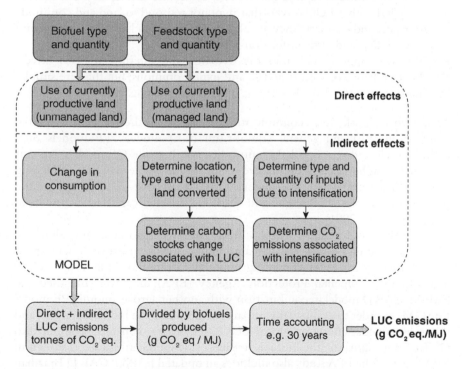

Figure 7.1 A schematic description of LUC determination methodology.

Source: Adapted from Rosenfeld and Pont (2010).

With increasing demand for biofuels, both dLUC and iLUC can occur at the same time, and often the iLUC of one industry is the dLUC of another. When agricultural or pasture land previously destined for the food, feed and fibre markets is diverted to the production of biofuels, the non-fuel demand will still need to be satisfied. Although this additional demand can be met through intensification of the original production, bringing non-agricultural land into production elsewhere is also possible. It is in the latter case that LUC is considered to have occurred indirectly. Whereas the LUC effect is considered 'indirect' to biofuels production, it is a 'direct' LUC for the food, feed and fibre markets occurring as a consequent of increasing demand for biofuels. However, most models do not distinguish between direct and indirect LUCs and report the overall GHG emissions as LUC factors.

Current models to assess LUC

To assess LUC and LUC factors, many models have been developed; however, these existing models report a wide range of iLUC factors representing significant uncertainties about the results. Basically, these iLUC models can be grouped into economic and non-economic models, as follows.

- Economic models capture market and economic drivers to LUC but fail to include other LUC drivers that may not respond to land and commodity prices and elasticity; they are also complex, often lack transparency and accessibility, and difficult for non-experts to use.
- Non-economic models miss the dynamic and interlinked LUC drivers due to reliance on historical data and expert opinions; but they are simple, transparent and relatively easy for stakeholders to engage.

As shown in Table 7.1, economic models can be further divided to partial equilibrium (PE) and general equilibrium (GE) models. PE models only include a selected set of traded commodity goods while GE models account for interactions between all productive sectors in the whole economy, including factor markets (in which the materials or factors that are essential to the production process are bought and sold) for labour, capital and land (van Tongeren et al., 2001).

AGLINK-Cosimo and FAPRI-CARD are two agro-economic dynamic PE models. They only model the agricultural sector and treat bioenergy as an additional demand for certain agricultural commodities (de Vries, 2009). The US Environmental Protection Agency (EPA), however, has used the FAPRI-CARD model in combination with another agro-economic PE model FASOM, in order to evaluate the impacts of US policy on biofuels demand and the consequential effects on global LUCs. These are then used as input into GREET for estimating the biofuels net life-cycle GHG emissions, as shown in EPA (2010). The EPA study also includes an updated FAPRI-CARD Brazilian model (Brazilian Land Use Model, BLUM), which is based on improvements to Brazilian land use data.

Table 7.1 Summary of characteristics, strengths and limitations of existing LUC models

| | Economic models | Non-economic models | | |
	PE and GE	Causal-descriptive	Deterministic method	System dynamics
Examples	AGLINK-Cosimo (dynamic PE) FAPRI-CARD (dynamic PE) GTAP (static GE) MIRAGE (dynamic GE) GCAM (integrated PE)	Bauen et al. (2010) Nassar et al. (2010), a Brazil-specific model Schmidt et al. (2012), a Life-Cycle Assessment method	Tipper et al. (2009)	BioLUC
Description	Calculate LUC based on marginal changes resulting from shocking a market system at equilibrium with biofuel expansion	Use cause-and-effect relationships to assess market response resulting from biofuel expansion	Allocate LUC GHG emissions to biofuels based on the proportion of commercial agricultural products considered relevant to the bioenergy sector	Use a dynamic stock-and-flow system to assess market response resulting from biofuel expansion
Data	Historical data used for calibration but may incorporate empirically estimated functions; model parameters are often from literature	Use historical trends as inputs and for calibrating cause-and-effect relationships	Based on historical land use data	Use historical trends as inputs and for calibrating cause-and-effect relationships
Strengths	Capture market and economic drivers to LUC; in their dynamic forms can account for evolution and changes over time	Can incorporate LUC drivers other than economic factors; can be transparent and easily used by stakeholders	Easy to calculate and transparent	Can incorporate LUC drivers other than economic factors; provides insights into the drivers and interactions of LUC; can be transparent and easily used by stakeholders

(continued)

Table 7.1 *(continued)*

	Economic models	Non-economic models		
	PE and GE	*Causal-descriptive*	*Deterministic method*	*System dynamics*
Limitations	Currently fail to include other LUC drivers that may not respond to land and commodity prices and elasticity, for example, policy and regulations Difficult to access and use by non-experts in economic modelling	Miss the dynamic and interlinked LUC drivers due to heavily reliance on historical data Data and relationships are at a relatively aggregated low-resolution level Price is not explicitly modelled	Miss the market feedback and policy measures Does not consider land types not used for producing traded agricultural commodities Data and relationships are at very aggregated low-resolution level Price is not explicitly modelled	Miss the dynamic and interlinked LUC drivers due to reliance on historical data Data and relationships are at a relatively aggregated low-resolution level Price is not explicitly modelled Lack of inter-regional dynamics
References	Davies (2012), Tyner et al. (2010), EPA (2010), Laborde (2011), PNNL (2012)	Bauen et al. (2010), Nassar et al. (2010), Schmidt et al. (2012)	Tipper et al. (2009)	Warner et al. (2013)

Source: Adapted from Warner et al. (2013).

In contrast to these agro-economic models, GCAM is an integration of a PE global energy model, a LUC model and two climate models. Rather than being treated as part of agricultural sector as in AGLINK and FAPRI-CARD, bioenergy in GCAM is modelled as a share in total energy demand that is driven by economic and population growth (Kyle et al., 2011).

The initial GE model GTAP, used by California's Air Resources Board (CARB) in the Low Carbon Fuel Standard (LCFS), has been modified to produce two extended versions, GTAP-BIO and GTAP-AEZ, by including improvements to the database, interactions between different agricultural sectors and the possibility of converting conservation reserve program (CRP) land in the USA (Rosenfeld and Pont, 2010). MIRAGE, which is another GE model, has been used by the International Food and Policy Research Institute (IFPRI) for the European Commission to assess the potential LUC impacts of the European Union's (EU's) biofuels mandates and trade policy.

In addition to these models, a few non-economic models have also been developed. These models generally estimate the LUC effects of biofuel feedstock by making assumptions about crop yields and displacement of by-products based on historical data/trends and expert opinions.

It is found that for economic models the uncertainties are primarily related to the underlying data and assumptions on the market-equilibrium parameters while for non-economic models assumptions made to overcome the lack of market interactions are the main sources of uncertainty. The uncertainties in the iLUC models can generally be grouped into three broad themes. They are:

- LUC-related uncertainties: this includes the effects of price increase on food consumption rate and crop yield as well as the assumed marginal crop yield and by-products use,
- land-allocation-related uncertainties: this relates to the methods and database used ('biophysical', 'historical' or 'economic' approach) within the model, as well as land supply elasticity and types of land cover,
- uncertainties related to GHG emissions calculations: this is associated with the carbon stock changes, time accounting method, foregone carbon sequestration and extra fertilizer use in intensification.

Models can be a helpful tool to assess iLUC and the associated GHG emissions (iLUC factors) resulting from increasing demand for biofuels. However, the selection of models and the methods for estimating the iLUC factors must be chosen carefully. From a more general perspective, it has been seen that:

- most models only focus on conventional biofuels except in the case of AGLINK and GCAM that have included advanced lignocellulosic biofuels;
- most PE models do not include the possibility of converting pasture to cropland. However, this is allowed for in the GE models through the use

of Constant Elasticity of Transformation (CET) functions (Tyner et al., 2010). Nevertheless, an accurate CET function to represent the competitiveness of pasture land relative to other types of land cover is difficult to calibrate with real data;

- most models have geographical preference due to the biofuel policy implanted and efforts on recruiting local data. For example, AGLINK, FAPRI-CARD and GTAP were used to study iLUC associated with biofuels expansion in the USA; IFPRI-MIRAGE and GTAP were used in EU studies; whereas the FAPRI-CARD Brazil module BLUM focused on Brazilian sugar cane ethanol expansion;

- for application purposes, economic models (PE and GE) are likely to be used to study the iLUC impacts induced by a biofuel 'shock' (additional demand) and have been used to examine the effects of biofuel policy (by comparing iLUC impacts with and without biofuel policy), deterministic models [the methods of Schmidt et al. (2012) and Nassar et al. (2010) based on actual data are used to generate the historical iLUC impacts caused by biofuels expansion in the past], dynamic models (e.g. BioLUC is preferred to study the drivers to iLUC factors) and predictive scenarios (e.g. food and biofuel demand scenarios in the future). As an emerging new approach, dynamic system models for LUC will be described in more detail below.

System dynamics for modelling LUC

A system dynamics model can be built around a specific problem, by using stocks and flows, to simulate real-world systems to capture nonlinearities, mutual interaction, information feedback and circular causality. System dynamics modelling is an emerging approach to assessing LUC and a model called BioLUC has been built recently by the National Renewable Energy Laboratory (NREL) to study LUC associated with increases in biofuel demand (Warner et al., 2013). Another model based on system dynamics is the Global Calculator project, which includes global simulations for land use, food and bioenergy. Both models are briefly described in the following sections.

BioLUC model

BioLUC was developed as a global land use simulation model using the Stella™ software package that provides an internal model logic based upon a system dynamics modelling approach, combined with a visual programming environment (Richmond, 1985). The most recent BioLUC model, version 10 (Bryant, 2013), describes trends in land use for 19 regions. All regions are modelled similarly, providing yearly accounts of the output. It also models resolutions of internal animal product and crop stock supply/demand imbalances through import and export activities (Warner et al., 2013).

The land cover in BioLUC is classified as 'Forest and Grassland, Pasture, Cropland and Abandoned land'. 'Forest and Grassland' includes mostly forest and also natural grassland while 'Cropland' is defined as land used to produce agricultural commodities and 'Abandoned land' is defined as cropland that has degraded to the point that is no longer productive, or at least productive enough to be worthwhile to use (Bryant, 2013). The key input parameters such as regional population, per capita demand and crop/biomass yields are based on projections from the UN Food and Agriculture Organization (FAO), while biofuel demand is considered as an exogenous, policy-driven input parameter (Bryant, 2013). Warner et al. (2013) also show that land flows between land categories over time are driven by several key feedback processes: (1) imbalances between production and consumption of various agricultural products at a regional level such as crops for food, feed and fuel, (2) re-distribution of land bases, for example by converting available land into pasture or by turning pasture to agricultural land and (3) crop or animal product imbalances are further reduced through imports/exports from other regions.

Rather than predicting the future, BioLUC aims at providing insights into the drivers and dynamic interactions of LUC. In addition to its advantages, such as transparency and ease of use, scenarios can be easily developed to study various future conditions, test model integrity under high pressures and investigate different biofuel policies. For example, these scenarios include population scenarios, high food demand and/or high biofuel demand (driven by policy) scenarios, different yield (crops, animal products, and biomass for biofuel) scenarios, etc. More details and results regarding these scenario studies can be found in studies by Warner et al. (2013) and Bryant (2013). Besides these advantages of the BioLUC model, a few limitations are also indicated in these studies. These include not being able to examine land intensification, advanced biofuel systems that use wastes and residues, or growth of cellulosic fuel feedstocks on abandoned land. If the model considered these options LUC induced by biofuel demand increase could be assessed, providing another application with which to study potential LUC mitigation options.

The Global Calculator project as a tool for land use assessment

The Global Calculator[1] enables users to explore the options for reducing global GHG emissions associated with land, food and energy systems in the period to 2050. It builds on the success that a number of countries have had in developing their own country-level 2050 calculators but it will extend the approach by illustrating the detrimental impacts of climate change associated with global-level choices. The project is led by the UK Department of Energy and Climate Change (DECC), and co-funded by Climate-KIC, the Climate Knowledge and Innovation Centre of the European Institute of Innovation and Technology. Imperial College London is the sector lead for land use, food, bioenergy and GHG removal technologies.

The Global Calculator presents a novel methodological approach for modelling both carbon and land use dynamics at a global scale for the following sectors: transport; manufacturing; electricity; land, bioenergy and food; and buildings. It also considers climate change impacts, different rates of population growth and urbanization, and scenarios for the inclusion of speculative Greenhouse Gas Removal (GGR) technologies. All sectors and variables are interconnected in a dynamic model, which allows users to generate a large number of GHG emission reduction trajectories online. The Global Calculator can be used by decision makers in the public and private sectors to inform management strategies for carbon mitigation, LUC, food and biomass production.

The approach employed in the Land/Bio/Food module of the Global Calculator applies a mathematical model for balancing the necessary expansion in the production of food crops, livestock, biofuels and other bio-based products with resource conservation (Strapasson, 2014). It allows users to simulate a number of trajectories of LUC and its associated GHG emissions, according to different demands for land-dependent products and services by 2050. Users can then develop their preferred pathways to 2050 by varying the weight of a selected set of parameters ('Levers') according to their GHG mitigation objectives ('Levels' 1–4, with several intermediate levels, at increasing levels of ambition). These include:

- calories consumed per person,
- meat consumed (including different meat types),
- crop yields,
- livestock yields,
- bioenergy yields,
- bioenergy types, i.e. solid biomass and biofuels (biogas modelled in separate),
- use of surplus land, i.e. future land allocation, including forest dynamics and energy crops,
- wastes and residues,
- land use efficiency (i.e. multi-cropping effects, and integrated farming schemes, such as agro-forestry and agro-livestock systems).

The model also considers several additional variables for the calculations, including the use of fertilizers, agricultural losses, GHG emissions factors, feed conversion ratios, the proportion of animals raised in intensive production systems (feedlots), concentration of animals in grazing systems (i.e. more animals per hectare), limiting factors for land distribution, etc. The model also includes soil carbon dynamics and temporal adjustments. The accuracy of each trajectory is limited by the availability of and uncertainty associated with data for global-scale estimates and the restricted number of input parameters in the calculator, given the high complexity and uncertainty of all these levers. The model draws on several data sources, primarily FAO, International Energy Agency (IEA) and Intergovernmental Panel on Climate Change (IPCC) statistics, and representative international references on land use modelling, with the purpose of obtaining not only a robust and credible methodology, but also a simple and user-friendly calculator for the lay user.

The Global Calculator is presented as a web tool, which was built on a database generated by a C language program (Ruby) from a comprehensive model in Microsoft Excel format. The model has several input parameters and variables, which are used for estimating future land use distributions, as well as the associated carbon dioxide, nitrouzs oxide and methane emissions. LUC is determined by a hierarchy of land use types. Priority is given to food production (crop and pasture lands) and the remaining land area is allocated to forestation, natural regeneration and/or energy crops. Figure 7.2 presents the driver tree of the Land/Bio/Food methodology.

Thus, with the Global Calculator it is possible to simulate a large number of trajectories for food, bioenergy and forest land by 2050, and as a result also assess the respective land use potential for sustainable biomass, according to the user's choice. The land use dynamics presented by the calculator includes both dLUC and iLUC, given that it is based on a global balance of several land use allocations. Figure 7.3 illustrates the dynamics of bioenergy, wastes and residues in the calculator.

In addition to the models presented above, there are many mapping tools for identifying the favourable land for a sustainable biomass production, e.g. agro-ecological zonings, which can be used by any region or nation, as described in the following section.

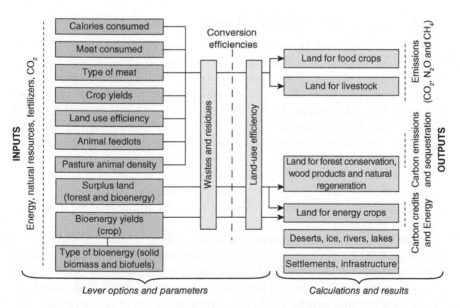

Figure 7.2 Driver tree for land, food and bioenergy at the Global Calculator model.

Source: Strapasson et al. (2014).

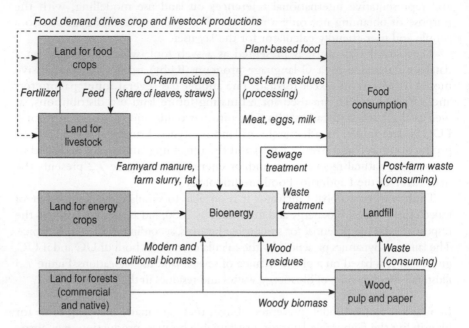

Figure 7.3 Dynamics of bioenergy, residues and wastes at the Global Calculator model

Source: Strapasson et al. (2014).

Land availability and zoning schemes

The FAO (2012) and Bruinsma (2011) estimated that about 30% of the world's land surface (4.5 billion ha) is suitable to some extent for rain-fed agriculture, according to Global Agro-Ecological Zoning (GAEZ) first published in 2002 by Fischer et al., as reported by Woods et al. (2014). Of this area, some 1.6 billion ha are already under cultivation (FAO, 2012). Developing countries have some 2.8 billion ha of land of varying qualities, which have potential for growing rain-fed crops at yields above an 'acceptable' minimum level, of which nearly 970 million ha are already under cultivation. The gross land balance of 2.9 billion ha (4.5−1.6 billion), 1.8 billion ha of which is in developing countries, would therefore seem to provide significant scope for further expansion of agriculture, including land for bioenergy feedstock provision.

In order to meet the projected increase in food demand globally, FAO (2012) projected a net increase in land demand by 2050 of 70 Mha resulting from an expansion of 130 Mha in developing countries and a decline of 60 Mha in developed countries. Therefore, at the global level, land constraints or competition may not be the key determinants of food or energy security, or the provision of biomass for bioenergy to date, although the dynamics for future projections are complex (FAO, 2012; Wise et al. 2014).

However, at the local level, increased demand for land for biofuels, as with other agricultural products, can stimulate LUC effects on local communities that range from positive, e.g. improved productivity and resilience, access to clean water and increased income (Satolo and Bacchi, 2013), to disastrous, e.g. loss of access to the land needed for subsistence, decreased access to water, increased food prices and detrimental impacts on biodiversity and ecosystem resilience. Understanding and mitigating these local-level constraints and impacts from the perspective of national and regional policy makers requires spatially and temporally explicit tools that can map local resources and enable a quantitative and dynamic classification of land for different uses, including the protection of biodiversity-rich or vulnerable land (see Box 7.1).

Box 7.1 Agro-ecological zoning for biofuels in Brazil

Agro-ecological zoning as a platform for sustainable biofuels expansion

As a result of local concerns and the need to provide strategic direction for the expansion of biofuel production in Brazil from sugar cane, a pioneering effort was made to develop a national agro-ecological zoning (AEZ) tool and supportive legislation. Brazil's 'Sugar cane Agroecological Zoning' was published as a legal framework in 2009 and involved a variety of legislation [MAPA Decree (Portaria) no. 333/2007, Presidential Decree no. 6961/2009, MAPA Normative Instruction no. 57/2009, Federal Law Project no. 6077/2009], with the work being co-ordinated by the Ministry of Agriculture, Livestock and Food Supply (MAPA et al., 2009). Key aims of the zoning are to direct sugar cane expansion onto existing managed lands, mainly pastures, avoid forest areas and avoid direct competition with food production (Figure 7.4). It also allows for the implementation and monitoring of reserve land in production areas such as buffer strips and biodiversity corridors as stipulated in Brazil's revised Forestry Code.

The sugar cane AEZ was based on satellite images and *in loco* analysis carried out by Embrapa and other partner institutions. The following areas were excluded from sugar cane expansion:

- Amazon and 'Pantanal' (swampland) biomes,
- lands with any kind of native vegetation (including 'Cerrado' and Atlantic Forest),
- lands without soil and climate favourable conditions,
- lands that would require full irrigation for producing sugar cane,

(continued)

(continued)

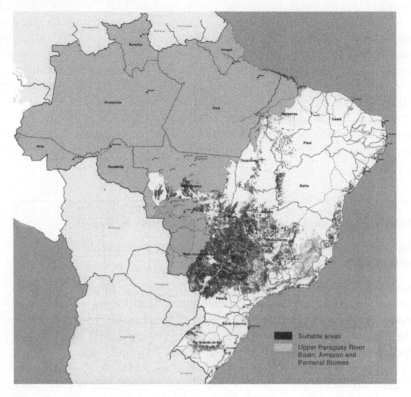

Figure 7.4 Sugar cane AEZ in Brazil.

Source: MAPA et al. (2009).

- topography with a slope greater than 12%, to allow mechanical harvest,
- protected areas and indigenous reserves,
- lands with high conservation value for biodiversity.

Since its publication in 2009, all the new sugar cane industries have been installed in full compliance with such zoning schemes. Therefore, AEZ has acted not only as a precautionary environmental measure, but also as a platform for stimulating sustainable investments. The suitable areas account for 64 million ha, which is much more than the current 8 million ha used for sugar cane, but still relatively small compared to the total Brazilian territory (851 million ha). Sugar cane zoning has been followed by zoning for oil palm (Figure 7.5), and Brazil has carried out workshops in a number of African countries to support capacity development in those countries for AEZ (Strapasson et al., 2012).

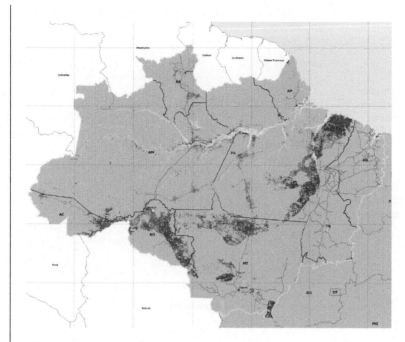

Figure 7.5 Oil palm AEZ in Brazil.

Source: MAPA et al. (2010). Shaded areas are all suitable areas for oil palm.

Fundamental steps needed to assess land use

Land use assessment for sustainable biomass is a strategic issue for developing bioenergy programmes. Such assessment is subject to the land use dynamics of agriculture. Figure 7.6 shows the main variables connected to LUC, which need to be addressed when estimating biomass potentials.

Although there is a great deal of speculation and concern regarding potential iLUC effects, the agricultural dynamics are so complex that the available model outputs continue to provide highly uncertain and inaccurate information to estimate such effects, as demonstrated in a comparative analysis by Akhurst et al. (2011) and Langeveld et al. (2014), and cited by Woods et al. (2014). Thus, it is largely unfeasible to quantify, monitor and control iLUC, especially in the current globalized food market (IPCC, 2013; Strapasson, 2014). However, precautionary measures via the implementation of specific public policies could reduce and even avoid iLUC effects, for example by using energy crops with higher yields, expanding biofuels onto underused or marginal areas, using AEZs and sustainable public policies for guiding new investments into favourable lands, and promoting capacity building and technology transfer in developing countries (Strapasson et al., 2012; Lynd and Woods, 2011).

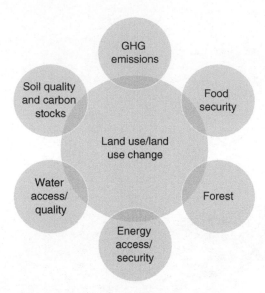

Figure 7.6 Main issues associated with land use and LUC.

Therefore, in order to prepare land use assessment at the regional or national level, the following steps should be followed when starting a project, as a basic guidance. For further discussions, see Strapasson (2014).

- Identify the key institutions responsible for land use data and territorial planning in the target area, such as governmental agencies, local authorities and research institutions.
- Gather all the useful land use data possibly available for the target region, such as surveys, maps and Geographic Information Systems (GIS) images. This includes maps that show the location of agricultural and pasture fields, native ecosystems, infrastructure and topography, as well as soil, temperature and precipitation maps. If not available, or if the database is not sufficiently robust, then new data should be prepared. Mapping information should also be checked in loco whenever necessary to ensure the accuracy of the database.
- Identify the current bioenergy areas (if any) and areas with land use potential for sustainable biomass production, for example solid biomass (e.g. eucalyptus, pine tree and other woody plants for cooking, heating or power) or biofuels crops (e.g. sugar cane, oil palm, oilseed rape). Selected crops should have positive net energy balances and net carbon savings when compared to the equivalent fossil fuel options (e.g. sugar cane ethanol vs petrol, rapeseed biodiesel vs mineral diesel). It is also important to obtain specific data on the environmental impacts of the production of the chosen crop.

- Assess the ongoing landscape and watershed planning beyond the crop of interest in order to design a sustainable system, where the chosen energy crop is a major but not the unique one to have a planned system.
- Exclude sensitive areas, in full compliance to environmental law, and consult stakeholders and local communities about the land availability initially identified. Ensure that the selected areas for energy biomass will not damage the local food supply as well.
- Then estimate the total biomass that could be produced in the selected lands, under a Life-Cycle Assessment (LCA), based on the type of crop selected for the project and its agronomical yields and cycles for the target region. The effective biomass potential would also depend on the economic feasibility for its expansion. In order to estimate the potential dLUC and iLUC impacts some references cited in this chapter may also be useful. Impacts on the biomass processing (e.g. biofuels plant) should follow a specific environmental impact assessment (EIA).

The credibility of the assessment is highly associated with the accuracy of the database, the methodology adopted and stakeholder involvement. Besides, projects may vary case by case, depending on local circumstances, legal framework and, in some cases, even political will. Funding agencies, if any, may also request to cover additional issues in the project, e.g. impacts on number of jobs and the infrastructure required for meeting a certain land use potential for bioenergy. The steps above are just a simple guidance on some of the key issues for a biomass assessment. A comprehensive assessment would have many more dimensions. For more information about multi-dimensional approaches for biomass assessments, and the limits of bioenergy, see Strapasson (2014).

Thus, there are a number of issues to look at when developing a land use assessment project, demanding professionals with complementary backgrounds and expertise in the area. To ensure the environmental, social and economic benefits of a bioenergy project it is essential to understand the complexity behind rural development, food security and forest conservation issues. It is certainly a multi-disciplinary task, which relies on systems thinking as a basis for a comprehensive assessment.

Note

1 For more information about the Global Calculator, including its methodology, spreadsheet and supporting papers, as well as the option to run the model, see www.globalcalculator.org.

References

AKHURST, M., KALAS, N. AND WOODS, J. (2011) Synthesis of European Commission biofuels land use modelling. *Science Insights for Biofuel Policy* 1, March 2011.
BAUEN, A., CHUDZIAK, C., VAD, K. AND WATSON, P. (2010) *A causal descriptive approach to modelling the GHG emissions associated with the indirect land use impacts of biofuels.* A study for the UK Department for Transport. Technical report published by E4tech, October 2010. London.

BRUINSMA, J. (2011) *The resource outlook to 2050: by how much do land, water and crop yields need to increase by 2050?* Paper presented to Expert Meeting on How to Feed the World in 2050, 24–26 June 2009. Food and Agriculture Organization of the United Nations (FAO), Economic and Social Development Department, Rome.

BRYANT, E. (2013) *BioLUC model parameters: comparison of time series to existing literature and sensitivity analysis.* National Renewable Energy Laboratory (NREL), Golden, CO.

DAVIES, G. (2012) *Removing biofuel support policies: an assessment of projected impacts on global agricultural markets using the AGLINK-COSIMO model.* Department for Environment, Food and Rural Affairs (DEFRA), London.

DE VRIES, J. (2009) *Exploring bioenergy's indirect effects: economic modelling approaches.* 2eco prepared for Netherland Environmental Assessment Agency.

EPA (2010) *Renewable fuel standard program (RFS2) regulatory impact analysis.* Technical report, reference no. EPA-420-R-10-006, February 2010. Assessments and Standards Division, Office of Transportation and Air Quality, US Environmental Protection Agency, Washington DC.

FAO (2012) *World agriculture towards 2030/2050: the 2012 revision*, Alexandratos, N. and Bruinsma, J. ESA Working Paper no. 12-03. UN Food and Agriculture Organization, Rome.

IPCC (2013) *Climate effects, mitigation options, potential and sustainability implications: appendix bioenergy (final draft) to chapter 11 (AFOLU).* IPCC, Geneva.

KYLE, P., LUCKOW, P., CALVIN, K., EMANUEL, W., NATHAN, M. AND ZHOU, Y. (2011) *GCAM 3.0 agriculture and land use: data sources and methods.* Technical report, reference no. PNNL-21025, December 2011. Pacific Northwest National Laboratory, USA

LABORDE, D. (2011) *Assessing the land use change consequences of European biofuel policies.* Final report, contract no. TRADE/07/A2, October 2011. International Food Policy Institute (IFPRI), ATLASS Consortium. DG-Trade, European Commission, Brussels.

LANGEVELD, J.W.A. ET AL. (2014) Biofuels, analyzing the effect of biofuel expansion on land use in major producing countries: evidence of increased multiple cropping. *Biofuels, Bioproducts and Biorefining* 8(1), 48–58.

LYND, L.R. AND WOODS, J. (2011) Perspective: a new hope for Africa. *Nature* 474, S20–S21.

MAPA, EMBRAPA, CONAB, MMA, CASA CIVIL, MDA ET AL. (2009) *Sugarcane agro-ecological zoning*, Strapasson, A.B., Caldas, C.J., Manzatto, C.V. et al. (National Coordination Group). Documents include: Presidential Decree 6.961/2009; Portaria MAPA Ministerial Decree n. 333/2007 and Normative Instruction 57/2009; Bacen Resolution 3.813/2009 and Resolution 3.814/2009; Federal Law Project 6.077/2009; Booklet and institutional video. Maps and GIS database available at the Embrapa Soils Geoportal. Brasília, Brazil.

MAPA, EMBRAPA, CASA CIVil, MMA, MDA ET AL. (2010) *Palm oil agro-ecological zoning.* Published by the Brazilian Government as part of the Sustainable Oil Palm Production Program, Campello, T., Brandão, S.M.C., Strapasson, A.B., Ferreira, D., Manzatto, C.V., Ramalho-Filho, A. et al. (National Coordination Group). Documents include: Presidential Decree 7.172/2010, Bacen Resolution 3.852/2010, Federal Law Project 7.326/2010; booklet and institutional video. Maps and GIS database available on the Embrapa Soils Geoportal. Brasília, Brazil.

NASSAR, A.M., ANTONIAZZI, L.B., MOREIRA, M.R., CHIODI, L. AND HARFUCH, L. (2010) *An allocation methodology to assess GHG emissions associated with land use change.* Institution for International Trade Negotiation (ICONE), São Paulo, Brazil.

PNNL (Pacific Northwest National Laboratory) (2012) *Agriculture, land use, and bioenergy.* PNNL, US Department of Energy. http://wiki.umd.edu/gcam/index.php?title=Agriculture,_Land-Use,_and_Bioenergy.

RICHMOND, B. (1985) Stella: software for bringing system dynamics to the other 98%. In *Proceedings of the 1985 international conference of the system dynamics society: 1985 international system dynamics conference.* Systems Dynamics Society, Keystone, CO., pp. 706–718.

ROSENFELD, J. AND PONT, J. (2010) *Indirect land use change and comparative analysis.* TIAX LLC, Cupertino, CA.

SATOLO, L.F. AND BACCHI, M.R.P. (2013) Impacts of the recent expansion of the sugarcane sector on municipal per capita income in São Paulo State. *ISRN Economics* 2013, article ID 828169.

SCHMIDT, J.H., REINHARD, J. AND WEIDEMA, B.P. (2012) *Modelling of indirect land use change in LCA, report v3.* 2.-0 LCA Consultants, Aalborg.

STRAPASSON, A.B. (2014) *The limits of bioenergy: a complex systems approach for land use dynamics and constraints.* PhD thesis, Imperial College London.

STRAPASSON, A.B., KALAS, N. AND WOODS, J. (2014) *Briefing paper on land, food and bioenergy of the global calculator project.* www.globalcalculator.org.

STRAPASSON, A.B., RAMALHO-FILHO, A., FERREIRA, D., VIEIRA, J.N.S. AND JOB, L.C.M.A. (2012) Agro-ecological zoning and biofuels: the Brazilian experience and the potential application in Africa. In *Bioenergy for sustainable development and international competitiveness: the role of sugar cane in Africa,* Johnson, F.X. and Seebaluck, V. (eds). Routledge, Earthscan Book, New York, Published simultaneously in the USA and Canada, pp. 48–65.

TIPPER, R., HUTCHISON, C. AND BRANDER, M. (2009) *A practical approach for policies to address GHG emissions from indirect land use change associated with biofuels.* Technical paper, reference no. TP-080212-A, January 2009. Ecometrica and Greenenergy, London.

TYNER, W., TAHERIPOUR, F., ZHUANG, Q., BIRUR, D. AND BALDOS, U. (2010) *Land use changes and consequent CO_2 emissions due to US corn ethanol production: a comprehensive analysis.* Center for Global Trade Analysis, Purdue University, West Lafayette, IN.

UN (United Nations) (2011) *UN 2010 projections.* Population Estimates and Projections Section, Population Division, Department of Economics and Social Affairs, United Nations, New York.

VAN TONGEREN, F., VAN MEIJL, H. AND SURRY, Y. (2001) Global models applied to agricultural and trade policies: a review and assessment. *Agricultural Economics* 26, 149–172.

WARNER, E., INMAN, D., KUNSTMAN, B., BUSH, B., VIMMERSTEDT, L., PETERSON, S. ET AL. (2013) Modelling biofuel expansion effects on land use change dynamics. *Environmental Research Letters* 8, 015003, number 1.

WISE, M., DOOLEY, J., LUCKOW, P., CALVIN, K. AND KYLE, P. (2014) Agriculture, land use, energy and carbon emission impacts of global biofuel mandates to mid-century. *Applied Energy* 114, 763–773.

WOODS, J., STRAPASSON, A., RAVINDRANATH, N.H. AND RACK, M. (2014) *Biofuels for climate change mitigation: update for the Global Environmental Facility (GEF).* Technical report, GEF, Washington DC (in press).

8 Remote sensing for mapping, monitoring vegetation dynamics and providing biomass production estimates

Jansle Vieira Rocha and Rubens A.C. Lamparelli

Introduction

In the previous edition of this handbook the focus of remote sensing was on forestry. To the contrary, this chapter focuses on agriculture, e.g. image analysis and applications in land use mapping and monitoring. Remote sensing techniques are becoming increasingly important in understanding global land and resource uses and as a tool for improving the management of land including the production of food, monitoring deforestation and managing environmental impacts of land use change (LUC). The increasing use of biomass for bioenergy is related to recent dynamics in land cover, with possible impacts on environment and climate. Examples and case studies are presented, mainly for LUC related to bioenergy crops, such as sugar cane, in Brazil and Africa.

Remote sensing is the science (Fussel et al., 1986; Jensen, 2006) that allows the gathering of information about a target without being in direct contact with it by analysing its interaction with electromagnetic radiation. In this case 'target' means any component of Earth's surface or sub-surface. There are different definitions for 'remote sensing' in the literature, with Showengerdt (2007), defining remote sensing as the measurement of object properties on the Earth's surface using data acquired from aircraft and satellites. Definitions may vary according to the context, so we assume that the basic principle of remote sensing is the analysis of electromagnetic radiation across different wavelengths and its interaction with targets on Earth's surface.

Physical principles and the electromagnetic spectrum

Electromagnetic radiation reaches and interacts with the target according to its characteristics and the wavelength. The radiation generated after this interaction with the target brings information that differentiates it from other targets allowing its identification and study. The spectral characterization of a target is based on analysis of the response of the target to radiation along the electromagnetic spectrum, the so-called spectral curves (Figure 8.1).

Vegetation, soil and water spectral curves

Considering the Earth and the vast range of materials that form it, we have an almost infinitely large number of targets with different characteristics.

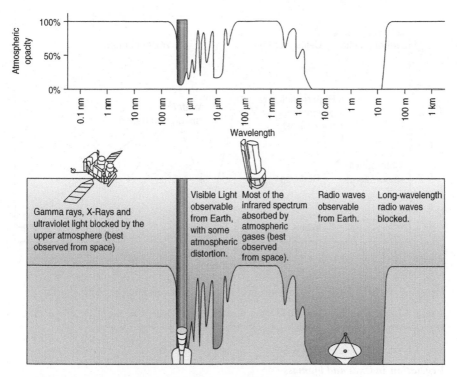

Figure 8.1 Atmospheric windows permeable to electromagnetic radiation.

Source: http://commons.wikimedia.org/wiki/File:Atmospheric_electromagnetic_transmittance_or_opacity.jpg.

To simplify this it is usual to characterize the targets that are the most observed in remote sensing into three broad categories: vegetation, soil and water. There is a great diversity of characteristics within these targets but it is assumed that each target presents a general form of spectral behaviour that can be used to distinguish it from other features, as shown in Figure 8.2.

For monitoring vegetation dynamics there are three regions of the electromagnetic spectrum in which different factors control the spectral response to radiation. In the shorter wavelengths, within the visible spectrum (blue, green and red), the photosynthetic pigments (chlorophyll, xanthophylls and carotene) found in plant tissue influence the response. In the near infrared region the internal structure of plant tissues, which is related to biomass, influence the spectral response. The variation of the spectral response that occurs in the middle infrared is basically influenced by the presence of water in plants, as shown in the absorption peaks.

In general, organic matter, texture (sandy, silty and clay), water content and colour are factors that influence spectral response of soils whereas sediments (siliceous, aluminum and iron oxide), chlorophyll and organic matter influence the spectral response of water.

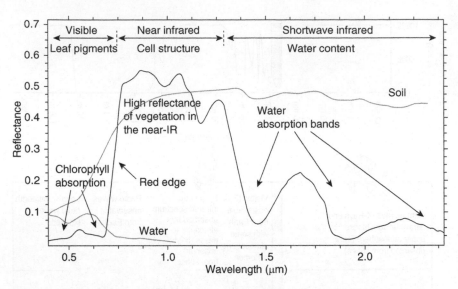

Figure 8.2 General spectral behaviour of some common targets on the Earth's surface.

Source: Adapted from the Science Education through Earth Observation for High Schools (SEOS) Project, www.seos-project.eu/modules/classification/classification-c00-p05.html.

Vegetation indexes and biomass

Vegetation indexes (VI) were created to maximize information brought by each spectral band, aggregating the positive effects that aid extraction and modelling of biophysical parameters. According to Jensen (2006), VIs are non-dimensional radiometric measurements that indicate the relative abundance and activity of green vegetation, including Leaf Area Index (LAI), percentage of green cover, chlorophyll content, green biomass and the fraction of absorbed photosynthetically active radiation (fAPAR). There are several VIs outlined in the literature (see References and further reading) that can be applied to monitoring plant growth. Table 8.1 provides some examples. The Normalized Difference Vegetation Index (NDVI) and the Enhanced Vegetation Index (EVI) are the most used for vegetation and biomass monitoring.

Sensors

Sensors are equipment, aboard satellite platforms, that are capable of transforming radiation received into an electronic signal that can be stored as digital information and then processed and converted into digital greyscale images. These images have specific characteristics, called resolutions.

Table 8.1 Description of some existing VIs

VI	Equation	Reference
Ratio (R)	$R = \rho_{nir} / \rho_{red}$	Birth and MacVey (1968)
Normalized Difference Vegetation Index (NDVI)	$NDVI = \rho_{nir} - \rho_{red} / \rho_{nir} + \rho_{red}$	Rouse et al. (1974)
Water Index	$NDWI = \dfrac{NIR_{TM4} - MidIR_{TM5}}{NIR_{TM4} + MidIR_{TM5}}$	Gao (1996)
Soil Adjusted Vegetation Index (SAVI)	$SAVI = \dfrac{(1-L)(\rho_{nir} - \rho_{red})}{\rho_{nir} + \rho_{red} + L}$	Huete (1988)
Enhanced Vegetation Index (EVI)	$EVI = G\dfrac{\rho_{nir}^{\star} - \rho_{red}^{\star}}{\rho_{nir}^{\star} + C_1\rho_{red}^{\star} - C_2\rho_{blue}^{\star} + L}(1 + L)$	Huete et al. (1997)

Spatial, spectral, radiometric and temporal resolution

Remote sensing systems register information in images with four different resolutions: spatial, temporal, radiometric and spectral.

- *Spatial resolution* is the ability of the sensor to 'see' and detect a target of a minimum size. This value is defined by the pixel size, which is derived from the sensor's instantaneous field of view (IFOV). It may vary from centimetres to kilometres, depending on the sensor and its destined application. The greater the spatial resolution the greater will be the ability to detect small targets.
- *Temporal resolution* is given by the time interval between two sequential image acquisitions of a specific place on the surface. It can be hours or days and the resolution increases as the sensor observes a target at smaller intervals of time.
- *Radiometric resolution* is the ability of a sensor to generate images with different grey levels. It is expressed as a base 2 number (2^x), where x is the number of bits (e.g. 8 bits will represent an image with 2^8 grey levels, i.e. 256 grey levels). The bigger the number of bits, the greater will be the radiometric resolution.
- *Spectral resolution* is related to the number of spectral bands, or intervals of wavelengths, registered by the sensor. The bigger the number of bands, the greater the spectral resolution of the sensor.

Main sensors

There are several sensors available aboard satellite platforms that are able to perform agricultural/environmental monitoring. The USA and France have a long tradition of developing Earth observation satellites. Two of the most successful satellite programmes, the US Landsat Land Satellite and the French Satellites pour l'Observation de la Terre (SPOT), have been operational since the 1970s and 1980s. Since the launch of Landsat 1, the Earth Resources Technology Satellite (ERTS), which is still operational, a further seven satellites have been put into orbit in the Landsat series, with the latest, Landsat 8, launched in 2012. France launched its first satellite (SPOT1) in 1986. In 2012, SPOT6 was launched, and SPOT7 was launched in June 2014. Some of the main satellite/sensors available are shown in Table 8.2.

The last 10 years have seen the development of a new generation of sensors, with spectral, radiometric, spatial and temporal resolutions appropriate to agricultural and environmental monitoring, such as Terra/Aqua, RapidEye and Spot/Pleiades. These sensors, with more and narrower bands, facilitate the identification of vegetation components and an increased availability of radar data, can be used for continuous monitoring even under cloud cover.

Vegetation mapping and monitoring

Vegetation monitoring, especially crop monitoring, can be used to help prevent/minimize environmental effects, such as diseases, water, nutrient and temperature stresses. Mapping and monitoring vegetation over a long period of time also allows area estimates and land use and land cover change analysis (Araújo et al., 2011), providing an important tool for environmental impact assessment (Adami et al., 2012).

Classification techniques

Lu and Weng (2007) revised some image processing techniques, encompassing several aspects involved in extracting information from images. They considered that the major steps of image classification may include determination of a suitable classification system, selection of training samples, image preprocessing, feature extraction, selection of suitable classification approaches, post-classification processing and accuracy assessment. The user's needs, scale of the study area, economic conditions, analyst's skills, the design of the classification procedure and the quality of the classification results are important factors influencing the selection of remotely sensed data.

Multi-temporal profiles of VIs

The NDVI and EVI, the most widely used VIs, give information on the responses of plants to their environment, expressed in terms of biomass (e.g. kg/m^2). The relationship between these indexes, as well as leaf area and plant biomass (Tucker, 1979; Jackson and Huete, 1991), are direct, and rapid changes in their parameters (or no change over time) could indicate problems with crop vegetative growth cycles. Each type of vegetation has its own characteristic behaviour, due to

Table 8.2 Main satellite/sensors in use

Sensors	Satellite	Country	Spectral resolution (bands)	Radiometric/spatial resolution (SR)/revisit time (RT)
OLI (Operational Land Imager)	Landsat 8[a]	NASA (National Aeronautics and Space Administration), USA	Band 1: 433–453 nm Band 2: 450–515 nm Band 3: 525–600 nm Band 4: 630–680 nm Band 5: 845–885 nm Band 6: 1,560–1,660 nm Band 7: 2,100–2,300 nm Band 8: 500–680 nm Band 9: 1,360–1,390 nm	12 bits/SR 15–30–100 m/RT 16 days
TIRS (Thermal Infrared Sensor)			Band 10: 10,600–11,200 nm Band 11: 11,500–12,500 nm	
NAOMI (New AstroSat Optical Modular Instrument)	Spot 6 and 7[b]	CNES (Centre National d'Études Spatiales)	Panchromatic: 450–745 nm Blue: 450–520 nm Green: 530–590 nm Red: 625–695 nm Near IR: 760–890 nm	12 bits/SR 2–8 m/RT 1–5 days, depending on off-nadir angle
SAR (Synthetic Aperture Radar)	Radar sat 1 and 2[c]	CSA (Canadian Space Agency), Canada	C Band – HH (Radarsat 1) C Band - Full polarimetric (Radarsat 2)	12 bits/SR 3–100 m/RT 3, 7 or 24 days, depending on off-nadir angle
SAR	TerraSAR-X[c]	DLR (Deutsches Zentrum für Luft- und Raumfahrt), Germany	X Band	16 bits/SR 1,3 or 16 m/RT 11 days at nadir and 2.5 days depending on off-nadir angle

(continued)

Table 8.2 (continued)

Sensors	Satellite	Country	Spectral resolution (bands)	Radiometric/spatial resolution (SR)/revisit time (RT)
SAR	TanDEM-X[c]	DLR, Germany	X Band	16 bits/SR 1–18 m/RT 11 days at nadir and 2.5 days depending on off-nadir angle
	RapidEye[d]	BlackBridge, Germany	Blue: 440–510 nm Green: 520–590 nm Red: 630–690 nm Red edge: 690–730 nm Near IR: 760–880 nm	12 bits/SR 6.5 m/RT 1 day
	Ikonos[e]	GeoEye, USA	Pan: 405–1,053 nm Blue: 430–545 nm Green: 466–620 nm Red: 590–710 nm Near IR: 715–918 nm	11 bits/SR 0.82–3.2 m (at nadir)/RT 5 days off-nadir and 144 days at nadir angle
AWiFIS (Advanced Wide Field Sensor) LISS 3 (Linear Imaging Self Scanning Sensor) LISS 4	IRS-P6[f] (Indian Remote Sensing)	ISRO (Indian Space Research Organization), India	Green: 520–590 nm Red: 620–680 nm Near IR: 770–860 nm Medium IR: 1,550–1,700 nm	10 bits/SR 56, 23.5 and 5.8 m/RT 5 or 24 days (depending of sensor and off-nadir angle)
	Quickbird 2[g]	DigitalGlobe, USA	Pan: 450–900 nm Blue: 450–520 nm Green: 520–600 nm Red: 630–690 nm Infrared: 760–900 nm	11 bits/SR 0.62 – 2.6 m/RT 1 – 3.5 days

Pleiades 1	Pleiades 1 and 2[h]	CNES, France	Pan: 480–820 nm, Blue: 450–530 nm, Green: 510–590 nm, Red: 620–700 nm, Infrared: 775–915 nm	12 bits/SR <1 m/RT 1–2 days
Pleiades 2				
MODIS (Moderate-resolution Imaging Spectroradiometer)		NASA (National Aeronautics Space Administration), USA	MODIS 36 bands (620–2,155 nm)	12 bits/SR 250–500–1,000 m/RT 1–2 day (s)
Aster (Advanced Spaceborne Thermal Emission and Reflection Radiometer)	Terra/Aqua[i]		14 bands (520–11,650 nm)	8 or 12 bits/SR 15, 30 or 90 m/RT 16 days
Forthcoming projects				
EnMap (Environmental Mapping and Analysis Programme)		DLR, Germany	250 bands (420–1,000 nm, VNIR; 900–2,450 nm, SWIR)	≥14 bits/SR 30 m/RT 4 to 21 days (depending to off-nadir angle)

Notes

[a]http://landsat.gsfc.nasa.gov/wp-content/uploads/2012/12/20101119_LDCMbrochure.pdf;

[b]https://directory.eoportal.org/web/eoportal/satellite-missions/s/spot-6-7;

[c]https://directory.eoportal.org/web/eoportal/satellite-missions/r/radarsat-2;

[d]www.blackbridge.com/rapideye/about/satellites.htm;

[e]www.digitalglobe.com/sites/default/files/DigitalGlobe_Spectral_Response_0.pdf;

[f]http://calval.cr.usgs.gov/documents/IRSP6.pdf;

[g]https://directory.eoportal.org/web/eoportal/satellite-missions/q/quickbird-2;

[h]http://smsc.cnes.fr/PLEIADES/GP_satellite.htm;

[i]http://asterweb.jpl.nasa.gov/characteristics.asp.

variation in plant genetics and the response of each species to environmental and biophysical parameters. It is therefore possible to identify and classify vegetation types and their growth parameters and use this information to analyse LUC by developing temporal profiles of VIs (Arvor et al., 2012; Brown et al., 2013).

Figure 8.3 shows the temporal behaviour of different targets based on NDVI temporal series data from the MODIS/Terra sensor. Clear differences between targets can be identified over a 1 year monitoring period.

Knowledge of a 'typical' VI temporal profile for each type of vegetation cover allows mapping techniques to be developed based on these profiles as well as detection and mapping of land cover changes over longer periods (Bruzzone and Smits, 2002; Carrão et al., 2008) (Figures 8.4 and 8.5) and even biomass productivity and stocking estimates (Fernandes et al., 2011).

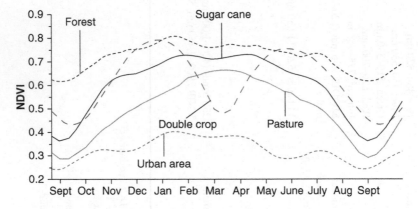

Figure 8.3 Temporal NDVI profiles for different targets.

Source: Modified from Fernandes (2009).

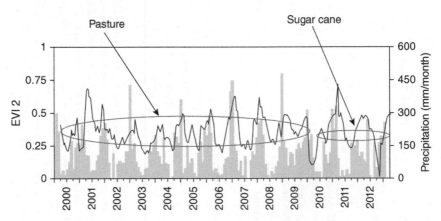

Figure 8.4 Land cover change from pasture to sugar cane with temporal profile of VI. Grey bars indicate precipitation.

Source: Produced by the authors from www.dsr.inpe.br/laf/series.

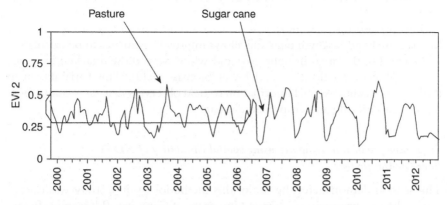

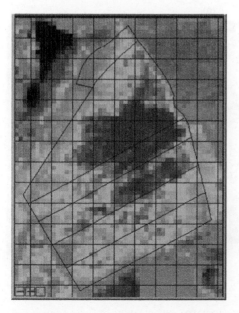

Figure 8.5 Land cover change from pasture to sugar cane.

Source: Produced by the authors from www.dsr.inpe.br/laf/series.

Monitoring and mapping biomass growth with satellite imagery

VIs from medium- to low-resolution satellites can also be used to develop biomass stocking and yield estimates (Fernandes et al., 2011). Figure 8.6 is an NDVI image generated from red (630–690 nm) and near infrared (760–900 nm) spectral bands of the Landsat 7 ETM+ (Enhanced Thematic Mapper) sensor.

Figure 8.6 Sugar cane within-field NDVI variability derived from Landsat 7 ETM+. For details see text.

Source: Produced by the authors at the Geoprocessing Lab, Unicamp, Brazil.

The colour differences, from darker grey to light grey and white, indicate variation from higher to lower biomass stocks (Table 8.3).

This processing of remotely sensed data helps to identify differences in biomass stocks and growth rates and allows improved estimates to be calculated. In Figure 8.6, the areas in light grey and white were those attacked by pests, causing photosynthetic stress and lower biomass yields. This variability map therefore provides essential information to field observers that carry out in-field biomass monitoring and crop yield estimates.

Sugar cane production estimates using spatial variability of NDVI from Landsat images

This section demonstrates the use of this methodology in a study was carried out by the Geoprocessing Group at University of Campinas (Unicamp), Brazil, with a major sugar mill in Brazil. The study compared traditional methods of production and yield estimates, based on visual field observations and some supplementary biomass measurements (e.g. plant height) by specialized technicians, with Landsat NDVI variability images (Figure 8.7). The traditional methods for estimating crop yields are normally associated with estimation errors up to 10%.

The NDVI images were used to select specific points, located with geographical coordinates, where, using GPS, the technicians carried out traditional visual

(a)   (b)

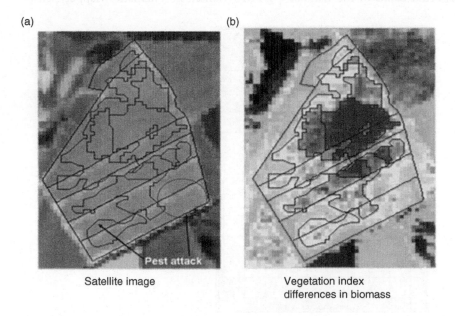

Satellite image   Vegetation index
                                   differences in biomass

Figure 8.7 (a) Satellite image; (b) classified NDVI with variability/biomass differences.

Source: Authors, prepared at the Geoprocessing Lab, Unicamp, Brazil.

Table 8.3 Estimated yield and production values for 2001 by traditional sugar mill evaluation and NDVI methods proposed by Unicamp

Field	Area (ha)	Production (kg)							Yield (t/ha)		
		Actual	Mill estimates			Unicamp/NDVI			Actual	Mill	Unicamp
			kg	Difference from actual	% Difference	kg	Difference from actual	% Difference			
1	22.07	1,621	1,766	144	9%	1,600	-22	-1%	73	80	72
2	22.93	1,556	1,834	279	18%	1,703	147	9%	68	80	74
3	24.49	2,012	1,959	-53	-3%	2,121	110	5%	82	80	87
4	18.72	1,550	1,591	41	3%	1,605	55	4%	83	85	86
5	53.19	4,518	4,521	3	0%	4,580	62	1%	85	85	86
6	7.31	530	585	55	10%	452	-78	-15%	72	80	62
Total		11,787	12,256	469	4%	12,061	274	2%			

estimates and measurements of biomass stocks. These points represented places with low, intermediate and higher biomass, according to the NDVI difference (Figure 8.7b).

New general estimates (Figure 8.8b) were generated based on field measurements and compared to the original sugar mill estimates, based only on field observations. Table 8.3 on page 239 shows the comparison, for different fields, between the original estimates and the new estimates, using NDVI variability maps, reducing the error from 4% (original) to 2% (with NDVI).

LUC detection using time series of MODIS Terra/Aqua images

Using the analysis of VI temporal profiles it is possible to distinguish between different land cover types (Lamparelli et al., 2012; Mello et al., 2013). Figure 8.9 shows two typical curves for sugar cane (Figure 8.9a), which is a semi-perennial crop, and an annual crop/double-cropping system (Figure 8.9b). The parameters associated with each specific curve (e.g. minimum, maximum, difference, slope coefficient) allow the characterization and identification of each land cover for equatorial to tropical/sub-tropical regions. In (a) Δa means difference between the maximum and minimum NDVI values during the whole growing cycle and Δb means difference between the maximum and minimum values during the maximum growing period. In (b) Δa means difference

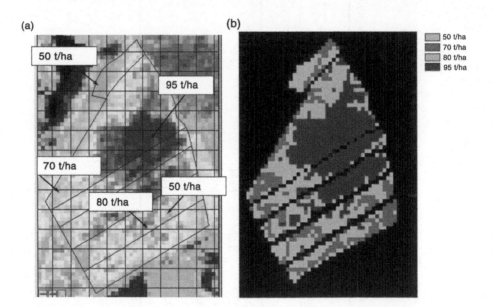

Figure 8.8 (a) Points evaluated by the technicians; (b) new estimates based on biomass measurements.

Source: Produced by the authors at the Geoprocessing Lab, Unicamp, Brazil.

(a)

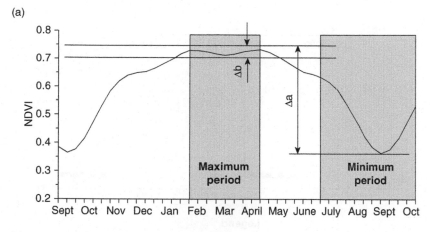

(b)

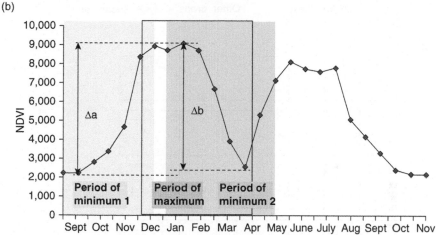

Figure 8.9 (a) Parameters of NDVI profiles for sugar cane; (b) annual double cropping. For details see text

Source: Prepared by the authors at the Geoprocessing Lab, Unicamp, Brazil.

between the maximum and minimum NDVI values during the initial growing cycle and Δb means difference between the maximum and minimum NDVI values during the final growing cycle.

Once the temporal profile and the parameters are defined, it is possible to generate maps (masks) for different land cover and, depending on the environment and crop calendar characteristics, it is possible to generate general cropland masks, or even crop type masks. The long-period analyses of these cropland/crop type masks provide an essential input to land use and LUC mapping and evaluation (Galford et al., 2008, 2010). Figures 8.10 and 8.11

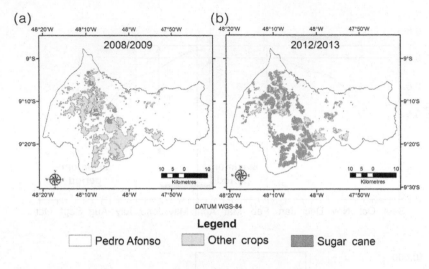

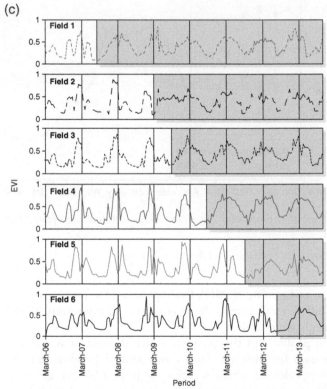

Figure 8.10 LUC as result of transition from pasture to soybean (2000–2007) and from soybean to sugar cane (from 2010) in Pedro Afonso, Tocantins State, Brazil. For details see text.

Source: Produced by the authors at the Geoprocessing Lab, Unicamp, Brazil.

(a)

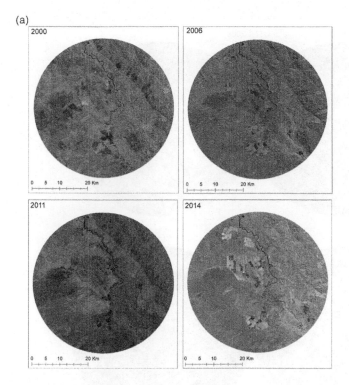

(b)

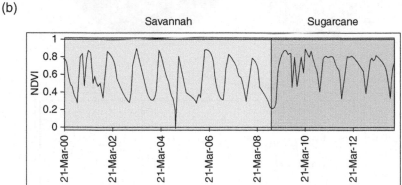

Figure 8.11 LUC as result of transition from cerrado (Brazilian savannah) to irrigated sugar cane in Itacarambi, Minas Gerais State, Brazil. For details see text

Source: Produced by the authors at the Geoprocessing Lab, Unicamp, Brazil.

(Brazil) and Figure 8.12 (Africa) show satellite images of the same areas taken in different periods. Figure 8.10a shows the LUC map between 2009 and 2013, panel b shows NDVI profiles of the LUC areas (2000–2013) and panel c shows NDVI profiles of different periods of change (2006–2013) in specific fields. Figures 8.11 and 8.12 show LUC over a period of 14 years (2000–2014)

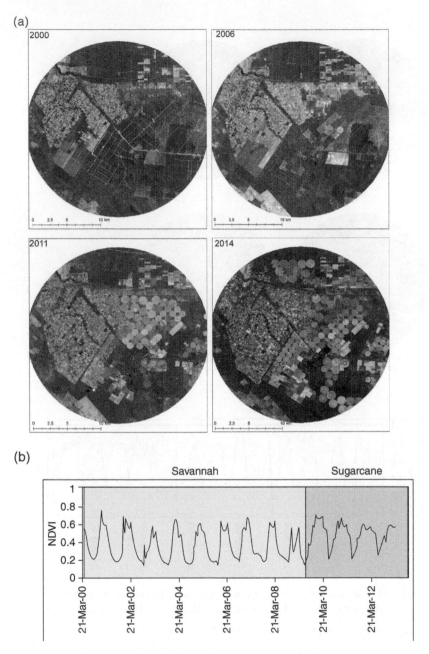

Figure 8.12 LUC in Malawi as a result of the expansion of sugar cane over savannah.
For details see text

Source: Produced by the authors at the Geoprocessing Lab, Unicamp, Brazil.

as a result of sugar cane expansion in central Brazil (Figure 8.11) and Africa (Malawi; Figure 8.12). These images clearly show the LUCs occurring in these regions and are associated with their respective temporal VI profiles, where it is possible to detect when and where LUC has occurred as well as the previous land cover/ cropland/crop type.

Potential use of remote sensing for carbon stocks assessment

Recent studies have emphasized the importance of LUC, particularly when estimating soil carbon stocks (Anderson-Teixeira et al., 2013; Berenger et al., 2014; Mello at al., 2014). Mello et al. (2014) studied the effects of LUC on soil carbon (C) balance as a result of a transition to sugar cane in Brazil, calculating payback times for recovering the carbon lost to the atmosphere through LUC, the so-called 'soil carbon debt'. Szákacs et al. (2011) used remote sensing to study soil carbon stocks under pasturelands and concluded that LAI, derived from orbital remote sensing, is suitable for estimating soil organic carbon (SOC) at regional to global scales. While not discussed here, remote sensing is also now widely used to monitor deforestation and reforestation at regional to global scales and is playing an important role in understanding the greenhouse gas impacts of both deforestation and reforestation and the size of the carbon sink that currently occurs on land.

Suchenwirth et al. (2012) used simulations based on land use maps generated by digital classification of satellite images to estimate the total carbon storage of soils and vegetation (crops, grassland and forests), and generated a map of their spatial distribution.

With the future availability of new hyperspectral and radar sensors aboard satellite platforms, studies of LUC (Lambin and Geist, 2006; Lambin and Meyfroidt, 2011) as a result of the expansion of bioenergy/biomass crops will lead to more accurate global assessment of biomass mapping, production and environmental impact evaluation. This will definitely contribute to global policies on land use and land cover, balancing the need to fulfill the growing demand for food with an increase in the use of bioenergy.

References and further reading

ADAMI, M.M., RUDORFF, B.F.T., FREITAS, R.M., AGUIAR, D.A. ET AL. (2012) Remote sensing time series to evaluate direct land use change of recent expanded sugarcane crop in Brazil. *Sustainability* 4(4), 574–585.

ANDERSON-TEIXEIRA, K.J., MASTERS, M.D., BLACK, C.K., ZERI, M. ET AL. (2013) Altered below ground carbon cycling following land use change to perennial energy crops. *Ecosystems* 16(3), 508–520.

ARAÚJO, G.K.D., ROCHA, J.V., LAMPARELLI, R.A.C. AND ROCHA, A.M. (2011) Mapping of summer crops in the state of Paraná, Brazil, through the 10day Spot Vegetation NDVI composites. *Engenharia Agrícola (Impresso)* 31, 760–770.

ARVOR, D., MEIRELLES, M., DUBREUIL V., BÉGUÉ, A. AND SHIMABUKURO, Y.E. (2012) Analyzing the agricultural transition in Mato Grosso, Brazil, using satellite-derived indices. *Applied Geography* 32, 702–713.

BERENGUER, E., FERREIRA, J., GARDNER, T.A., ARAGÃO, L.E.O.C., CAMARGO, P.B., CERRI, C.E., DURIGAN, M., OLIVEIRA JUNIOR, R.C., VIEIRA, I.C.G. AND BARLOW, J. (2014) A large-scale field assessment of carbon stocks in human-modified tropical forests. *Global Change Biology* 20(12), 3713–3726.

BIRTH, G.S. AND MACVEY, G. (1968) Measuring the color of growing turf which a reflectance spectrophotometer. *Agronomy Journal* 60, 640–643.

BROWN J.C., KASTENS, J.H., COUTINHO, A.C., DE CASTRO VICTORIA, D. AND BISHOP, C.R. (2013) Classifying multiyear agricultural land use data from Mato Grosso using time-series MODIS vegetation index data. *Remote Sensing of Environment* 130, 39–50.

BRUZZONE, L. AND SMITS, P. (2002) *Analysis of multi-temporal remote sensing images.* Series in Remote Sensing 2. World Scientific, Hackensack, NJ.

CAMPBELL, J.B. (2002) *Introduction to remote sensing*, 3rd edn. Taylor & Francis, New York.

CARRÃO, H., GONÇALVES, P. AND CAETANO, M. (2008) Contribution of multi-spectral and multitemporal information from MODIS images to land cover classification. *Remote Sensing of Environment* 112(3), 986–997.

FERNANDES, J.L. (2009) *Monitoring sugar cane crop plantations in São Paulo state using spot vegetation images and meteorological data.* MSc dissertation, Unicamp, Brazil. www.bibliotecadigital.unicamp.br/document/?down=000469139 [in Portuguese].

FERNANDES, J.L., ROCHA, J.V. AND LAMPARELLI, R.A.C. (2011) Sugarcane yield estimates using time series analysis of spot vegetation images. *Scientia Agrícola (USP. Impresso)* 68, 139–146.

FUSSEL, J., RUNDQUIST, D. AND HARRINGTON, J.A. (1986) On defining Remote Sensing. *Photogrammetric Engineering and Remote Sensing* 52(9), 1507–1511.

GALFORD, G.L., MELILLO, J., MUSTARD, J.F., CERRI, C.E.P. AND CERRI, C.C. (2010) The Amazon frontier of land-use change: croplands and consequences for greenhouse gas emissions. *Earth Interactions* 14(15), 1–24.

GALFORD, G.L., MUSTARD, J.F., MELILLO, J., GENDRIN, A., CERRI, C.C. AND CERRI, C.E.P. (2008) Wavelet analysis of MODIS time series to detect expansion and intensification of row-crop agriculture in Brazil. *Remote Sensing of Environment* 112, 576–587.

GAO, B.C. (1996) NDWI: a Normalized Difference Water Index for remote sensing of vegetation liquid water from space. *Remote Sensing of Environment* 58, 257–266.

HUETE, A.R. (1988) A Soil-adjusted Vegetation Index (SAVI). *Remote Sensing of Environment* 25, 295–309.

HUETE, A.R., LIU, H.Q., BATCHILY, K. AND VAN LEEUWEN, W.J. (1997) A comparison of vegetation indices over global set of TM images for EOS-MODIS. *Remote Sensing of Environment* 59, 440–451.

JACKSON, R.D. AND HUETE, A.R. (1991) Interpreting vegetation indices. *Preventive Veterinary Medicine* 11, 185–200.

JENSEN, J. (2006) *Remote sensing of environment: an Earth resources perspective*, 2nd edn. Prentice Hall, NJ.

LAMBIN, E.F. AND GEIST, H. (eds) (2006) *Land-use and land-cover change: local processes and global impacts.* Global Change series—The IGBP. Springer-Verlag, Berlin.

LAMBIN, E.F. AND MEYFROIDT, P. (2011) Global land use change, economic globalization, and the looming land scarcity. *Proceedings of the National Academy of Sciences USA* 108, 3465–3472.

LAMPARELLI, R.A.C., MERCANTE, E., ROCHA, J.V. AND URIBEOPAZO, M. (2012) Impact of the normalization process on the spectral-temporal profile of soybean crops (*Glycine max* (L) Merril) based on vegetation indexes. *International Journal of Remote Sensing* 33, 1605–1626.

LU, D. AND WENG, Q. (2007) Survey of image classification methods and techniques for improving classification performance. *International Journal of Remote Sensing* 28(5), 823–870.

MELLO, F.F.C., CERRI, C.E.P., DAVIES, C.A., HOLDBROOK, N.M., PAUSTIAN, K., MAIA, S.M.F., GALDOS, M.V., BERNOUX, M. AND CERRI, C.C. (2014) Payback time for soil carbon and sugar-cane Ethanol. *Nature Climate Change* 4, 1–5.

MELLO, M.P., VIEIRA, C.A.O., RUDORFF, B.F.T., APLIN, P.R.D., SANTOS, C. AND AGUIAR, D.A. (2013) STARS: a new method for multitemporal remote sensing. *IEEE Transactions on Geoscience and Remote Sensing* 51(4), 1897–1913.

ROUSE, J.W., HAAS, R.H., SCHELL, J.A. AND DEERING, D.W. (1974) Monitoring vegetation systems in the Great Plains with ERTS. In *Earth Resources Technology Satellite-1 Symposium, 3. Proceedings, vol. 1*. NASA, Washington DC, pp. 309–317.

SHOWENGERDT, R.A. (2007) *Remote sensing: models and methods for images processing*, 3rd edn. Academic Press, London.

SUCHENWIRTH, L., FÖRSTER, M., CIERJACKS, A., LANG, F. AND KLEINSCHMIT, B. (2012) Knowledge-based classification of remote sensing data for the estimation of below- and above-ground organic carbon stocks in riparian forests. *Wetlands Ecology and Management* 20(2), 151–163.

SZAKÁCS, G.G.J., CERRI, C.C., HERPIN, U. AND BENOUX, M. (2011) Assessing soil carbon stocks under pastures through orbital remote sensing. *Scientia Agricola* 68(5), 574–581.

TUCKER, C.J. (1979) Red and photographic infrared linear combinations for monitoring vegetation. *Remote Sensing of Environment* 8(2), 127–150.

9 Case studies

*With contributions from Frank Rosillo-Calle,
Sarah L. Hemstock, Vineet V. Chandra
and Teuleala Manuella-Morris*

Introduction

Case studies are used to illustrate step-by-step methods for calculating biomass resources and uses in modern applications, based on field work experiences, or to illustrate potentially perceived major changes or trends in a particular area, e.g. increasing use of bioenergy in modern applications. There are four case studies, individually authored, as detailed below.

- 9.1 International bioenergy trade, which examines the development of the international biotrade in biomass for energy and its wider implications; this is of course a new trend which could have major global impacts in supply and demand of biomass.
- 9.2 Deals with biogas use in small island communities; it is an illustration of small-scale use of traditional applications.
- 9.3 This case study explains in detail, step by step, how to develop biomass energy flow charts.
- 9.4 Deals with a major issue which shaping policy towards the use of biomass for energy in general and biofuels in particular: the 'food-versus-fuel debate'. It is important that all involved in biomass for energy are fully aware of the pros and cons of this debate.

Case study 9.1 International biotrade: potential implications for the development of biomass for energy

Frank Rosillo-Calle

Unlike fossil fuels, biomass for energy has hardly been traded regionally, nationally and, even less so, internationally. Most trading of solid biomass has occurred at local level, but this is now changing and it is rapidly becoming a major traded commodity, particularly as densified biomass (i.e. pellets and chips) and liquid biofuels (i.e. ethanol and biodiesel). To give a quick example, from 2000 to 2013 the growth of the pellet market was 23%, from 1.7 to 24.5 Mt; by 2020 this market is expected to reach 27 Mt; other estimates put it as high as 80 Mt by 2030. This new phenomenon could have major implications for the development of bioenergy, both the potential benefits and possible pitfalls. (See also Appendix 4.2, on densified biomass.)

The rapidly growing international biomass trade was recognized by the International Energy Agency (IEA), which subsequently created 'Bioenergy Task 40: sustainable international bioenergy trade: securing supply and demand', on which most of the information of this case study is based (see www.biotrade.org). Task 40 was established in December 2003 in Campinas, Brazil, and focuses on assessing the implications of international biotrade; it is playing a pivotal role in the development of biomass for energy at the international level.

The trade in biomass for energy is evolving rapidly and it is difficult to predict what shape the market will take. However, present trends indicate that there will be major developments, given growing international interest (political, technical and economic). For example, the European Union (EU) has various ongoing projects dealing with biomass for energy trade strategies (see www.biotrade2020plus.eu).

The rationale for bioenergy trade

Over the past decades, the modern use of biomass has increased rapidly in many parts of the world. Given the important role that biotrade is likely to play in the rapidly expanding international bioenergy sector both positive and negative outcomes are highly possible. It is important to be aware of any potential outcome.

In a world where the demand for energy seems endless and continuously changing, biomass for energy is bound to play a critical role. A reliable supply and demand is vital to develop stable market activities, aimed at bioenergy trade. Given the expectation of a high global demand for bioenergy, the pressure on available biomass resources will increase. Without the development of biomass resources (e.g. through energy crops, better use of agro-forestry residues, generation of new feedstocks and new conversion processes) and a well-functioning biomass market to ensure a reliable and lasting supply these ambitions may not be met; or, worse, they could have serious negative impacts on the sustainability of supply (see also Chapter 6).

The development of truly international markets may become an essential driver to develop the potential of bioenergy, which is currently under-utilized in many regions of the world. This is true for residues, dedicated biomass energy plantations (forestry and crops) and multi-functional systems such as agro-forestry. The possibility of exporting biomass-derived commodities for the world's energy market can provide a stable and reliable demand for rural communities, thus creating important socio-economic development incentives and market access. A key objective is to ensure that biomass is produced in a sustainable way and does not affect food production (see Case study 9.4).

Opportunities and pitfalls of bioenergy trade

The development of an international bioenergy trade will be unequal around the world because there are many regions with poor resources. For some developing countries endowed with large biomass resources, and lower costs, biotrade offers a real opportunity to export primarily to developed countries. Many developing countries have a large potential for technical agro-forestry residue as well as for dedicated energy plantations, e.g. ethanol from sugar cane and pellets or charcoal from eucalyptus plantations, and this could represent a good business opportunity.

International biotrade is still very small, but growing rapidly. Many trade flows are between neighbouring countries, but long-distance trade is also increasing. Examples are the export of ethanol from Brazil to Japan and the EU, palm kernel shells (a residue of the palm oil production process) from Malaysia to the Netherlands and wood pellets from Canada and the USA to the EU.

In the short to medium term, the most promising areas for international trading are (1) densified biomass (pellets, torrefied wood, wood chips) for combined heat and power and co-firing in coal power plants and small to medium-sized applications, (2) biofuels (biodiesel and ethanol for use in transport and (3) charcoal (i.e. use in restaurants), although this market will remain comparatively small.

Biofuels seem to be one of the most realistic options for setting up a truly global bioenergy trade, at least within the present decade. Ethanol fuel in particular is growing rapidly as it has considerable potential for substituting oil in the transportation sector, given the right conditions. The market for pellets is also growing even faster in some regions, such as the EU.

The international ethanol trade varies significantly from year to year. Between 3 and 5 billion litres of fuel ethanol are traded annually, with Brazil and the USA being the main exporters, and Japan and EU the main importers. But this is expected to increase dramatically as international demand is growing rapidly, with more than 40 countries interested in ethanol fuel. One of the major constraints is that few countries, excluding Brazil and a few others, are in a position to supply both their growing domestic market and international demand. For example, Japan, as well as other potential importers, would like to import far more but is constrained by the high inelasticity of the ethanol fuel market. If the ethanol fuel market becomes large enough it could create sufficient liquidity to attract many ethanol fuel players. A large ethanol market would be able to guarantee to cover the possibility of any shortfall in any one country or region, as is the case of oil or gas.

These trade flows may offer multiple benefits for both exporting and importing countries. For example, exporting countries may gain an interesting source of additional income and an increase in employment. Also, sustainable biomass production will contribute to the sustainable management of natural resources. Importing countries, on the other hand, may be able to fulfil cost-effectively their greenhouse gas (GHG) emission reduction targets and diversify their fuel mix.

For interested stakeholders, e.g. utilities, producers and suppliers of biomass for energy, it is important to have a clear understanding of the pros and cons of this market. Investment in infrastructure and conversion facilities requires risk minimization of supply disruptions, in terms of volume and quality as well as price.

The long-term future of the large-scale international biotrade must rely on environmentally sustainable production of biomass for energy. This requires the development of criteria, project guidelines and a certification system, all supported by international bodies. This is particularly relevant for markets that are highly dependent on consumer opinion, as is presently the case in the EU (see also Chapter 6).

What are the main drivers for international bioenergy trade?

Clear criteria and identification of promising possibilities and areas are of key interest as investments in infrastructure and conversion capacity rely on minimization of risks of supply disruptions. The following is a summary of the main emerging drivers in international biotrade (see www.biotrade.org).

- *Cost-effective GHG emission reduction*. At present, climate policies in various countries such as in the EU are major factors in the growing demand for bioenergy. In situations where indigenous resources are insufficient or costs are high, imports can be more attractive than exploiting local biomass potentials. This will change over time as costs will be reduced but several world regions will continue to have inherent advantages in producing lower cost of biomass-based fuels, particularly in tropical developing countries.
- *Socio-economic development*. There is ample evidence indicating that there is a strong positive link between developing bioenergy use and local (rural) development. Furthermore, for various countries bioenergy exports may provide substantial benefits for their trade balances.
- *Fuel supply security*. Biomass for energy will diversify the energy mix and thus prolong the lifespan of other fuels (fossil fuels) and reduce risks of energy supply disruptions. This argument is particularly strong in the case of liquid biofuels (ethanol and biodiesel) as the transport sector is overwhelmingly dependent on oil.
- *Sustainable management and use of natural resources*. Large-scale production and use of biomass for energy will involve, inevitably, additional demand on land. However, if biomass production can be combined with better agricultural methods, restoration of degraded and marginal lands, and improved management practices, to ensure it is produced sustainably, this in turn can provide a sustainable source of income for many rural communities (see Chapter 7).

Are there any particular barriers to international bioenergy trade?

Based on a literature review and interviews, a number of potential barrier categories have been identified. These barriers may vary a great deal in terms of scope, relevance for exporting and importing countries and how stakeholders perceive it, although this is constantly changing. A summary of the main barriers is given below.

Economic barriers

One of the principal barriers for the use of biomass energy in general is the competition with fossil fuel on a direct production-cost basis (excluding externalities). There are many hidden costs of fossil fuels which are not reflected in the market. Also, a stumbling block has been, and will remain, the high cost of feedstocks.

Many governments around the world have now introduced various mechanisms (legislative, subsidies, compulsory purchasing, etc.) to promote the use of biomass for energy, particularly for electricity, heat and mandatory blending in transportation (ethanol and biodiesel). However, often such support is not enough, particularly when it comes to long-term policies; this discourages long-term investment as bioenergy is still often considered too risky by many investors. This problem is further compounded by lack of harmonization at many levels, e.g. policy among EU Member States, standards, etc., are good examples.

Technical barriers caused by the physico-chemical characteristics of biomass

This is a major difficulty, as indicated throughout this handbook, which those dealing with bioenergy have to come to terms with. Whereas the market is gradually accepting and adapting to bioenergy, some major barriers remain. For example, physical and chemical properties such as low density, high ash and moisture content, nitrogen, sulphur or chlorine content, etc., make it more expensive to transport, and often unsuitable for direct use (e.g. in co-firing power plants). Overcoming many of these difficulties will require many technical improvements (e.g. boilers), and changes in the attitude of major users.

Logistical barriers

An important limiting factor to trading biomass is that it is often bulky and expensive to transport. One solution is to compact it to make it economic and practical to move long distances at acceptable cost. Fortunately, densification technology has improved significantly recently and, as detailed in Appendix 4.2, densified biomass is already being commercialized on a large scale. Liquid biofuels such as ethanol, vegetable oils and biodiesel do not present such difficulties and can easily be transported cheaply over long distances.

Large-scale transportation of bulky biomass is just beginning to take off and thus our experience is very limited. For example, very few ships exist today that are specially adapted for this purpose and this will probably result in higher transportation costs. However, some studies have shown that long-distance international transportation by ship is feasible in terms of energy use and transportation costs; but availability of suitable vessels and meteorological conditions (e.g. winter time in Scandinavia and Russia) need to be considered.

Harbours and terminals that do not have the capability to handle large biomass streams can also hinder the import and export of biomass to certain regions. The most favourable situation is when the end user has a facility close to the harbour, thus avoiding additional transport by trucks. The lack of significant volumes of biomass can also hamper logistics. In order to achieve low costs, large volumes need to be shipped on a more frequent basis.

International trade barriers

International biotrade is quite recent and thus often no specific biomass import regulations exist, which can be a major hindrance to trading. For example, in the EU most residues that contain traces of starches are considered potential animal fodder, and thus subject to EU import levies.

Although not exclusive to bioenergy, the potential contamination of imported biomass material with pathogens or pests (e.g. insects, fungi) can be a major impediment. For example, imported roundwood into EU can currently be rejected if it seems to be contaminated, as will be any other type of biomass.

Ecological barriers

Large-scale dedicated energy plantations also pose various ecological and environmental issues that cannot be ignored, ranging from monoculture, long-term sustainability, potential loss of biodiversity, soil erosion and water use, nutrient leaching and pollution from chemicals. However, let us keep a sense of reality: biomass for energy is no better or worse than any other traded commodity.

Social barriers

Large-scale energy plantations also pose potentially major social implications, both positive and negative; e.g. the effect on the quality of employment (which may increase, or decrease, depending on the level on mechanization, local conditions, etc.), the potential use of child labour, education and access to health care, etc. However, such implications will reflect prevailing situations and would not be better or worse than any other similar activity.

Land availability, deforestation and potential conflict with food production

This issue is covered in Chapter 7 and Case study 9.4. Food versus fuel is a very old issue that refuses to go away despite the fact that a large number of studies have demonstrated that land availability is not the real problem. Food security should not be affected by large energy plantations if proper management and policies are put in place; government policy and market forces would favour food production rather than fuel in the case of any food crisis. However, food availability is generally not the problem, but the lack of purchasing power of the poorer strata of the population to purchase it. In developed countries, where there is surplus land, the main issues may be competition with fodder production rather than food crops.

Methodological barriers: lack of clear international accounting rules

This can be a serious problem, at least in the short term, since clear rules and standards need to be established before international bioenergy trade can be developed. The nature of biomass can also pose problems for trade in biomass because it can be considered as a direct trade of fuels and as indirect flows of raw materials that end up as fuels in energy production during or after the production process of the main product. For example, in Finland the biggest international biomass trade volume is indirect trade of roundwood and wood chips. Roundwood is used as a raw material in timber or pulp production; wood chips are also raw material for pulp production. One of the waste products of the pulp and paper industry is black liquor, which is used for energy production.

Legal barriers

Biotrade is further complicated by the fact that each country, even countries within the EU, has its own legislation to deal with international trade. For example, emission standards differ a lot from country to country and this is an area that affects biotrade because of the potential for carbon credits, for example. Therefore common emission standards among major trading blocs will have important positive benefits for international biotrade.

So what lessons can be learned from bioenergy trade?

This case study has raised, rather than answered, a number of important issues. For example, some questions that need to be asked are: should biotrade be treated differently to any other commodity given its nature? Should market forces be left to determine supply and demand as any other commodity? Is biotrade so special that needs to be considered in its own right? How much control should there be of biotrade and will this enhance or hinder its development? Can international biotrade distort domestic supply? What does this mean potentially for traditional biomass applications?

Further reading

JUNGINGER, M., GOH, C.S. AND FAAIJ, A. (EDS) (2014) *International bioenergy trade: history, status and outlook on securing sustainable bioenergy supply, demand and markets.* Springer, Berlin.
www.bioenergytrade.org This site provides you with a lot of information on bioenergy trade (objectives of Task 40, workshops, documents, country reports, publications, contacts, etc.).

Case study 9.2   Biogas as a renewable technology option for small islands

Sarah L. Hemstock and Teuleala Manuella-Morris

Introduction

As indicated in Chapter 4, in the section on secondary fuels, biogas is an important source of energy, particularly in Asian countries such as China, India, Nepal and Vietnam. Although biogas is increasingly used in large industrial applications, its primary use is small scale. It is of particular interest for small rural communities where there are few other alternatives. Biogas is an attractive alternative because it can provide various simultaneous benefits: energy, sanitary waste disposal systems and fertilizer. This case study looks at the use of biogas in small island communities and focuses on the non-governmental organization (NGO) Alofa Tuvalu's successful Small is Beautiful (SiB) project.

Tuvalu is an independent constitutional monarchy in the south-west Pacific Ocean, located approximately 1,000 km north of Fiji with a total land mass of only 26 km² spread over 750,000 km² of its exclusive economic zone. It is composed of nine low-lying coral atoll islands and numerous islets, with the largest island covering only 520 ha and the smallest 42 ha. Average elevation is 3 m. The capital island, Funafuti, is only 2.8 km² but accounts for around half of the total population of approximately 11,000 and two-thirds of the gross domestic product (GDP) of the country.

The nation is regarded as exceptionally vulnerable to rising sea levels and increased storm activity as the maximum height above sea level is a mere 5 m. The climate is sub-tropical, with temperatures ranging 28 to 36°C uniformly throughout the year. There is no clear marked dry or wet season. The mean rainfall ranges between 2,700 and 3,500 mm per year but there are significant variations from island to island. Keeping pigs is a traditional activity and every household owns at least one pig (Rosillo-Calle et al., 2003; Woods et al., 2006; Alofa Tuvalu, 2005) (see Table 9.1).

Tuvalu is close to being a totally oil-dependent economy. In 2004 total energy consumption was 4.6 kilotonnes of oil equivalent (ktoe). Of this, oil accounted for 3.8 ktoe (82%) of total primary energy consumption and biomass accounted for 0.8 ktoe (almost 18% of total primary energy consumption) (also see Chapter 5, and Figures 5.1 and 5.2). This total includes the diesel charged by the two vessels (Nivanga II and Manu Folau) in Suva, Fiji (Hemstock and Raddane, 2005).

Table 9.1 Privately occupied households by island/region and number of livestock

Island/region	Number of households	Pigs	Chickens	Ducks	Cats	Dogs	Other
Funafuti	639	2,275	428	65	666	931	40
Outer islands	929	6,519	12,244	2,827	1,301	1,019	113
Tuvalu	**1,568**	**8,794**	**12,672**	**2,892**	**1,967**	**1,950**	**153**

Sources: Alofa Tuvalu (2005); Hemstock (2005); Tuvalu 2002 Population and Housing Census, Volume 2: Analytical Report; Social & Economic Wellbeing Survey 2003, Nimmo-Bell & Co.

Annual energy consumption

Annual energy consumption is over 0.4 tonnes of oil equivalent (toe) per capita (approximately one-third of someone living in Japan or Europe) (Hemstock and Raddane, 2005). Currently, all of Tuvalu's oil is imported. This is very vulnerable position since increasing oil prices and a drop in global oil production mean that Tuvalu's economy is in the hands of the oil suppliers.

Domestic kerosene

Domestic kerosene use in Tuvalu was estimated at 263 toe annually, which is considerable when compared with 170 toe for air transportation (Hemstock and Raddane, 2005).

The use of biogas as a fuel for cooking should reduce kerosene use. Production systems are relatively simple and can operate at small and large scales practically anywhere, with the gas produced being as versatile as natural gas. This is a very significant technology option for Tuvalu since anaerobic digestion can make a significant contribution to disposal of domestic and agricultural wastes and thereby alleviate the severe public health and water pollution problems that they can cause. The remaining sludge can then be used as a fertilizer (providing there is no polluting contamination) and actually performs better than the original manure since nitrogen is retained in a more useful form, weed seeds are destroyed and odours reduced. In addition, it would reduce the amount of household income spent on cooking fuel, provide much needed compost for family gardens (sea water encroachment has contaminated traditional taro pits) and thereby provide additional food security, dispose of pig waste (which is currently swilled into the lagoons) and may provide additional household income from pig rearing and sale of gas.

Additionally, the use of biogas systems in a country such as Tuvalu could increase agricultural productivity. All the agricultural residue, animal dung and human sewage generated in the community are available for anaerobic digestion. The slurry that is returned after methanogenesis is superior in terms of its nutrient content; the process of methane production serves to narrow the carbon:nitrogen ratio (C:N), while a fraction of the organic nitrogen is mineralized to ammonium (NH_4^+), and nitrate (NO_3^-), the form that is immediately available to plants. The resulting slurry doubles the short-term fertilizer effect of dung, but the long-term fertilizer effects are cut by half (Chanakya et al., 2005). However, in a tropical climate such as Tuvalu's the short-term effects are the most critical, as even the most slowly degrading manure fraction is quickly degraded due to rapid biological activity. Thus, an increase in land fertility could result in an increase in agricultural production. Value-added benefits include improved subsistence, increased local food security and income generation from increased land productivity.

Small-scale digesters

Small-scale digesters are appropriate for small and medium-sized rural farms. Floating-dome small-scale digester sizes range from 4 to 10 m³ total capacity and were built locally using plastic water tanks. Basic digester systems will produce around 0.5 m³ of biogas per m³ of digester volume. For a family of six, digester systems of 4–6 m³ in volume can meet daily biogas requirements (about 2.9 m³) for all residential and agricultural uses. For Tuvalu households, a 6 m³ floating dome was designed using two 'Rotamold' water tanks. Each unit was supplied with fencing for a 400 m² family garden, a 10 m³ pigpen with roofing, a rainwater-collection system and a 6 m³ water storage tank.

In theory biogas could be produced from the unused pig waste (Table 9.2). Using a conservative collection efficiency of 60% of total dung produced and conservative conversion efficiencies from family-based 6 m³ digesters (15 pigs per digester), a total of 1,578 m³ of gas could be produced per day, providing 13,236 GJ/year (315 toe). This is more than enough to provide the cooking gas (and possibly electricity for lighting) for 526 households across Tuvalu. Additional benefits include compost production, cleaner pig pens and the removal of a smelly hazardous waste.

Setting up the community initiatives

The 10 year SiB scheme was launched in 2005 via an initial inventory of available natural resources and renewable energy potential (Hemstock and Radanne, 2005)

Table 9.2 Energy available for biogas digestion from pig waste in Tuvalu

Total number of pigs in Tuvalu (head)	Total annual production of pig manure (tonnes/year)	Total annual production of energy from pig manure (GJ/year)	Total amount of manure available annually for use in biogas digester* (tonnes/year)	Total amount of energy available annually for use in biogas digester* (GJ/year)
12,328 (60% = 7,397)	3,600	32,400 (771 toe)	2,106	19,440 (463 toe)

*Assumes 60% collection efficiency as some waste is used for compost and some will be difficult to collect (Alofa Tuvalu, 2005; Hemstock, 2005).

and is aimed at addressing energy and related environmental problems associated with energy provision and resource use. Obviously, establishing a community need for energy services is the first step in the planning process, which was achieved via a series of community meetings. Initially community meetings were held with each of the eight island communities women's groups and included participatory assessments and technology sensitization. Failed projects show that communities must be involved in decision making and project planning from the outset (Woods et al., 2006). From the community meetings, training on all aspects of renewable energy technology (RET) installation and use was the most requested intervention, so a memorandum of understanding was drawn up between Alofa Tuvalu, an international NGO registered in France, with a counterpart in Tuvalu, and the Tuvalu Maritime Training Institute (TMTI) regarding the use of TMTI facilities as a RET Demonstration and Training Centre.

The basic format for each meeting was:

- to describe the background of Alofa Tuvalu and the SiB project,
- to make clear links between climate change, carbon emissions and energy use,
- to point out the benefits of family gardens (food security, improved income, reduced waste by compost making, reduced reliance on imported foods, etc.),
- to give instructions on how to set up a family garden (ground preparation, where to buy compost, planting seeds, watering, pollination, collecting and storing seeds, organic fertilizer and compost techniques, etc.),
- to supply seeds to the women (tomato, lettuce, basil, melon, marrow, chilli pepper),
- to describe biogas technology and types of implementation: examples of community biogas plants in India and a pig farm in Suva, Fiji; discussing added benefits of a reliable gas supply and compost production; providing women with information about the technology so that they can formulate an initial project strategy.

In general, the attendees were extremely positive about biogas and family gardens. The attendees requested help with compost making and marketing surplus home garden produce. However, there was some concern that Alofa Tuvalu was not operating through the correct channels. The women wanted to use the technology and thought the best way to implement a project would be through the Kaupule (local government).

Following community meetings which targeted women's groups and subsistence farmers, SiB's specific objectives were drawn up in response to community-identified needs. These included:

- reducing fossil fuel GHG emissions by improving energy efficiency and increasing renewable energy to a minimum 40% of primary energy consumption (4,574 toe),

- reducing poverty by improving access to affordable modern energy carriers and increasing household incomes (via production of coconut oil biodiesel and biogas-food initiatives),
- building capacity for technical and engineering support (ongoing training programmes in isolated outer-island communities on equipment use and maintenance).

Governance issues

Governance issues proved to be the key to the success of community biogas digesters in two outer-island communities in Tuvalu. Modernization of biomass energy use, via biogas, biodiesel and gasification, will involve some social and cultural changes. Political attitudes which emphasize the enforcement of existing policies will be required for successful implementation of the energy initiatives discussed.

The development of an RET Demonstration and Training Centre at TMTI was the initial starting point for the SiB initiative. TMTI is now equipped as a training centre for biogas, cookstoves (solar and charcoal), vegetable gardening (using biogas digester slurry), coconut biodiesel production, toddy bioethanol production and gasification. These technologies were chosen on the basis of: (1) robust and basic technology design – all could be manufactured and repaired in Tuvalu, (2) biomass is culturally accepted (traditional biomass energy use provides for the majority of domestic energy services), (3) biomass resources (coconuts and pigs) provide upstream and downstream income, and (4) resource availability – implementing replanting and reforestation as part of the SiB initiatives) (Hemstock, 2013).

TMTI was chosen for the basis for the RET centre since there are skilled marine engineers and classrooms and workshops available. TMTI is not attached to any one island community: it is a national asset. TMTI staff and students crew the inter-island boats so engineering staff and trainees could provide any technical and engineering backup for future RETs in the outer islands, ensuring technical sustainability on a national basis. Projects have failed in the past because, once the 'project cycle' ends, installed equipment breaks down and there are no parts or trained technicians to repair it. By partnering with TMTI, the SiB approach avoids these pitfalls.

As a RET training and demonstration centre, TMTI has been successful, training four people in the construction of 8 m fixed-dome digesters, 200 or more people on digester operation and maintenance (fixed and floating dome) and demonstrating gasification, toddy bioethanol and biodiesel (FuelPod) technologies to more than 500 people.

However, the use of the digester, gasifier and biodiesel Fuelpod to provide energy for TMTI has been less successful due to issues such as the reluctance of staff to use shared pig pens, vandalized equipment and lack of financial incentives since the Government of Tuvalu pays for TMTI's diesel for electricity production.

Implementation and local governance

Using the Nanumea community biogas installations as an example, the Nanumea Kaupule (local island council) and Alofa Tuvalu identified this project at an Island Council meeting in Funafuti in 2008. Nanumea was identified because it has the highest number of pigs per family (30 pigs per family) compared to other islands in Tuvalu, half the island is used for the rearing of pigs and there had been issues with pig waste, smell and large numbers of flies (thought to be due to the large amount of pig waste).

Alofa Tuvalu consulted on project design with the Nanumea Kaupule and the Nanumea community in Funafuti and secured funding from the United Nations Developmnent Programme Small Grants Scheme for the project. The Alofa Tuvalu consultant then visited Nanumea to discuss with the Kaupule and wider community the installation of biogas units, using lessons learned from the TMTI installation. A series of around 20 meetings (same format as above) were held with different groups within the community over a period of 1 month. Initially, the Kaupule and wider community (including students of Kaumaile Primary School) were introduced to the technology and those community members who had attended the TMTI digester training were encouraged to share their experiences.

From consultations with the Nanumea Kaupule it was decided that the Kaupule select four families to participate in this biogas project. This Kaupule decision was then taken to the Falekaupule Nanumea. The Falekaupule takes its name from the traditional meeting hall where all the elders and family heads, women and children gather for the general meeting with the Kaupule. These assemblies are held on a quarterly basis and it is here where annual local council budgets and plans are discussed and ratified. During these meetings the wider community get a chance to voice their views to the Kaupule on how projects are implemented and how they can be improved. One useful output for the project from this process was the recommendation that training the wider community in the construction, set up, operation and maintenance of biogas digesters should form a cornerstone of the project.

It was argued that if a large number of people in the community were trained, if the initial family selected to have a digester lost interest in the project then the digester could easily be moved to a different household (this has happened once since 2010). Recommendations were also made to improve sustainability of the project after the Alofa Tuvalu team handed over the infrastructure to the Kaupule. For example, TMTI recent graduates were selected to be on the team who would help the biogas consultant set up the digesters and train the wider community in operation and maintenance. Additionally, it was decided that the digesters were to be implemented on a household basis as this would reduce the arguments associated with community-based schemes, as demonstrated by the TMTI community digester.

Materials for the project (digesters, and materials for pig pens, freshwater storage/collection and family gardens) were shipped from Fiji to Funafuti, the capital of Tuvalu, and then on to Nanumea on the TMTI-operated boat Nivanga II. Alofa Tuvalu consulted Kaupule Nanumea as well as the Department of Rural Development in the Ministry of Home Affairs regarding the materials required for gardens and pig pens and shipping to Nanumea. This ensured national and local governance structures were engaged in the process.

In order to monitor the progress of the project on the outer island, Alofa Tuvalu employed a Project Coordinator, stationed on Nanumea, to assist the Kaupule in the implementation of the project and update Alofa Tuvalu on progress.

Problems

During 2011 severe drought took its toll throughout Tuvalu and a state of emergency was declared. The digestion process requires fresh water, but this was not available. However, due to the training received, the families responsible for the digesters used manure and green leaves to maintain some degree of gas production.

Stoves were replaced in December 2011 as the stoves supplied initially were not of good quality and did not survive Tuvalu's corrosive tropical environment.

Exit

In September 2012, Kaupule surveyed the four units and found out that all were well used. The feedback from the families with the units, especially from an elderly couple, was that the unit had changed their lives dramatically in the sense they no longer had to collect firewood and they were cooking 'smoke-less' inside their house. Refilling their unit only took two buckets of manure a week which was very easy for them to collect. Other families mentioned that the units had assisted them in cooking inside their main house during bad weather whereas some mentioned that they used their unit when there is a lack of imported gas on the island.

SiB installed the first outer-island biogas digesters in August 2010 (on a household rather than community basis), and they are still in daily use. Additionally, Tuvalu's outer island of Nanumaga has replicated the Nanumea project as part of the University of the South Pacific European Union Global Climate Change Alliance project by installing seven digester units, pigpens, water storage/collection systems and gardens at a household level, working through the Nanumaga Kaupule. This demonstrates that people are very able and willing to use alternative sources of energy which are local, convenient and renewable.

Conclusions

Although it has little impact on national fossil fuel use, the success of this low-cost project (US$50,000) for the community of Nanumea comes from using existing governance structures to identify a need for energy services, a thorough assessment of available resources and a rationale for implementation. This rationale allowed the participation of the wider community in the project, and promoted a transparent process for the selection of the beneficiary households/families by the wider community. In such manner, the families were publically committed to the project, knowing that they represented the community as a whole. This selection of the project sites and activities such as training by the wider community actually imposed the ownership of the project on them. In this case it would appear that the 'ownership' of the project by the wider community, even though the majority of benefits are at an individual household level, is the cornerstone of the project's success. This would not have been possible without the use of Nanumea's traditional governance structures.

Findings from Woods et al. (2006) are inherent in the design of the SiB project and account for capacity building via training and strengthening service provision, community involvement from project inception, appropriate technologies which can be manufactured and maintained without foreign agency 'technical assistance', the build-up of a critical mass of similar apparatus throughout Tuvalu, so systems maintenance is cost effective, and an integrated multi-disciplinary and multi-sector approach which builds on Tuvalu's existing infrastructure and institutions for service provision.

References

ALOFA TUVALU (2005) *Tuvalu field survey results: July–October 2005*. Alofa Tuvalu, Paris.
CHANAKYA, H.N., BHOGLE, S. AND ARUN, A.S. (2005) Field experience with leaf litter-based biogas plants. *Energy for Sustainable Development* IX(2), 49–62.
HEMSTOCK, S.L. (2005) *Biomass energy potential in Tuvalu*. Alofa Tuvalu, Government of Tuvalu.
HEMSTOCK, S.L. (2013) The potential of coconut toddy for use as a feedstock for bioethanol production in Tuvalu. *Biomass and Bioenergy* 49, 323–332.
HEMSTOCK, S.L. AND RADDANE, P. (2005) *Tuvalu renewable energy study: current energy use and potential for renewables*. Alofa Tuvalu, French Agency for Environment and Energy Management - ADEME), Government of Tuvalu.
MATAKIVITI, A. AND KUMAR, S.D. (2003) *Personal communication*. South Pacific Applied Geoscience Commission & Fiji Forestry Department of Energy, Suva, Fiji Islands.
ROSILLO-CALLE, F., WOODS, J. AND HEMSTOCK, S.L. (2003) *Biomass resource assessment, utilisation and management for six Pacific Island countries*. ICCEPT/EPMG, Imperial College London, SOPAC, South Pacific Applied Geoscience Commission.
WOODS, J., HEMSTOCK, S.L. AND BUNYEAT, W. (2006) Bio-energy systems at the community level in the South Pacific: impacts and monitoring. Greenhouse gas emissions and abrupt climate change: positive options and robust policy. *Journal of Mitigation and Adaptation Strategies for Global Change* 4, 473–499.

Websites

www.alofatuvalu.tv
www.youtube.com/user/AlofaTuvalu

Case study 9.3   A guide to the formulation and utilization of biomass energy flow charts

Vineet V. Chandra and Sarah L. Hemstock

Introduction

Flow charts are an effective tool to focus on a particular production stream and distribution across various systems. Hence biomass energy flow charts focus on biomass energy flows from different feedstocks and provide an understanding of the production, utilization and losses in biomass energy systems. This case study will discuss the methodology for deriving the biomass energy flow charts presented for Fiji by analysing two different periods (annual average data): 1999–2003 and 2008–2012.

Fiji is a small island nation consisting of approximately 330 islands situated between 177°E and 178°W longitude and 12° to 22°S latitude in the Pacific Ocean, spread over a land mass of 18,272 km². The largest island is Viti Levu, which covers 10,390 km², followed by Vanua Levu, which is 5,538 km². Biomass is one of the important sources of energy used in Fiji – particularly for domestic energy – and is extracted from feedstocks such as forestry, agricultural crop production and livestock.

The biomass energy flow charts for Fiji shown in Figures 9.1 and 9.2 were generated from data surrounding the three main areas of terrestrial biomass resources, namely forestry, agriculture, and livestock. Figure 9.1 provides energy flows (annual average) for years 1999–2003 and Figure 9.2 provides the same for years 2008–2012. The flow charts estimate total biomass energy theoretically available, its production, present consumption levels and the potential availability of biomass residues from forestry, agriculture and livestock. The flow of biomass in all its forms was traced from its production at source and harvest through to its end use and categorized into a product or 'end use' groups (e.g. food, fuel, residues, etc.). The following general assumptions apply to the flow charts in Figures 9.1 and 9.2.

All production refers to above-ground biomass; water surfaces were not considered as components of biomass energy production. The following energy values (GJ/tonne air dry, 20% moisture content) assume direct combustion was used (Hemstock and Hall, 1995, 1997; Amoo-Gottfried and Hall, 1999; Rosillo-Calle et al., 2006b):

- 1 tonne fuelwood = 15 GJ,
- 1 tonne 'roundwood' = 15 GJ,
- 1 tonne forest or tree harvesting residues (un-merchantable portion) = 15 GJ,
- 1 tonne charcoal = 31 GJ; efficiency of conversion for fuelwood to charcoal = 15% by weight,
- all 'roundwood' volume was assumed to be solid with a conversion equivalence of 1 m³ solid 'roundwood' = 1.3 tonne.

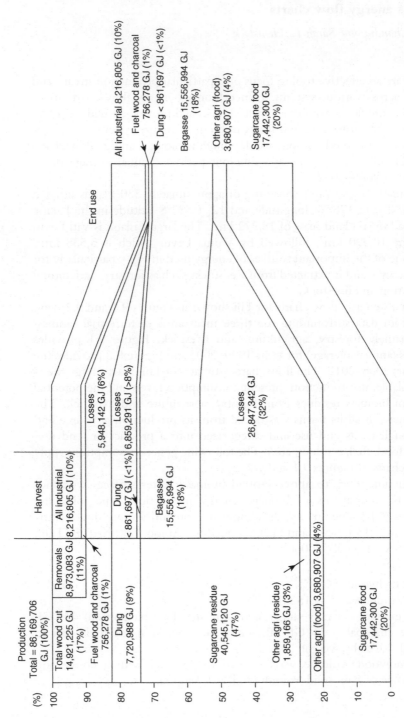

Figure 9.1 Biomass energy flow chart for Fiji (1999–2003). Based on annual average for 1999–2003 (FAOSTAT, www.fao.org). The scale is based on the total biomass harvested in agriculture and forestry. Figures in parenthesis indicate % total production. PJ = 10^15 J = 10^6 GJ. All industrial does not include miscellaneous wood residues [wood residues consist principally of industrial residues, e.g. sawmill rejects, slabs, edgings and trimmings, veneer log cores, veneer rejects, sawdust, bark (excluding briquettes), residues from carpentry and joinery production, etc.]. Total may not add up due to rounding.

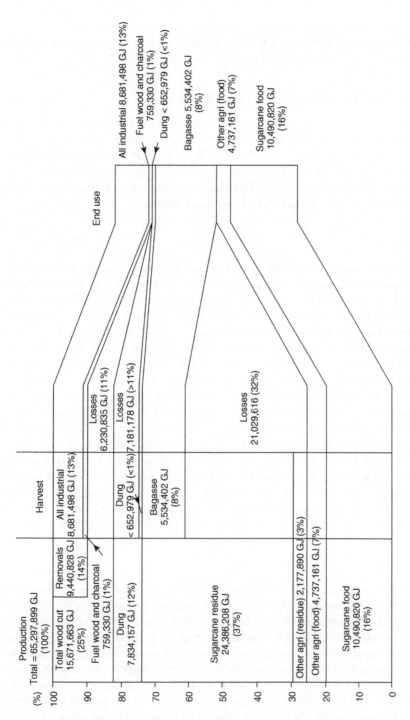

Figure 9.2 Biomass energy flow chart for Fiji (2008–2012). Based on annual average for 2008–2012 (FAOSTAT, www.fao.org). All details as for Figure 9.1.

Changing these assumptions to account for specific local/national/regional production will increase the overall accuracy of the flow chart. However, these assumptions were obtained from Hemstock and Hall (1995, 1997), Amoo-Gottfried and Hall (1999) and Rosillo-Calle et al. (2006b), where production characteristics were deemed to be similar to those of Fiji.

A methodology for drafting biomass energy flow charts

Terrestrial above-ground biomass production and utilization in Fiji was analysed for the years specified using FAOSTAT-derived data (FAOSTAT, 2014) (as biomass is seasonal and responds to variable environmental factors such as rainfall you should use data over an average of at least 3 years) as presented in Figures 9.1 and 9.2. The annual production of forestry, agriculture and livestock were recorded and analysed as follows.

Forestry

The forestry biomass contribution is measured by analysing the total 'round-wood' harvested from forest. Roundwood is wood which is in its natural state as felled, or harvested (wood which is usually transported to mills from forest after harvest). It is assumed that from the total wood cut 60% is 'roundwood' and 40% is residue that is left on site (Rosillo-Calle et al., 2006b). The annual production of 'roundwood' is obtained from available data (FAOSTAT) and using the conversion [Total wood cut (t) = 1.67 × 'roundwood' (t)] total wood cut was derived. Using an energy conversion coefficient (1 tonne = 15 GJ) the total energy from forestry (wood cut) can be derived. The annual production of 'fuelwood' was recorded and energy potential was derived using energy conversion coefficient (1 tonne = 15 GJ). 'Charcoal' production was recorded separately and the energy potential was derived using an energy conversion coefficient (1 tonne = 31 GJ). The category 'all industrial' wood does not contain any industrial residue and as such it only contains commercial timbers and products. However the value for 'total wood cut' accounts for the 60% 'roundwood' and 40% on-site forest residues assumed to be left *in situ* after harvest. Table 9.3 presents the results for forest biomass energy flows.

Crops

The different annual crop production values were obtained from Food and Agriculture Organization (FAO)-derived data for the years 1999–2003 and 2008–2012. The crop energy content, waste factors and residue energy contents are represented in Table 9.4. Annual average crop production, crop energy potential, annual average residue production and residue energy potentials are represented in Table 9.5. Land area used by different crops for the 5 year periods (1999–2003 and 2008–2012) is shown in Table 9.6. Table 9.4 indicates that groundnuts and coconuts are the most energy-dense crop products, followed by sorghum and maize. Sugar cane appears to be fairly low in terms of energy

Table 9.3 Summary of forest biomass energy production

Fiji		Units	Annual average (1999–2003)	Annual average (2008–2012)
Forestry total wood cut	Production (60% is roundwood)	m³	765,191	803,675
	Production (1.3 t/m³)	t	994,748	1,044,778
	Energy content (15 GJ/t)	PJ	14.92	15.67
Total biomass energy production		%	17	25
Forestry removals (use)	Production	m³	460,158	484,145
	Production (1.3 t/m³)	t	598,205	629,388
	Energy content (15 GJ/t)	PJ	8.97	9.44
Total biomass energy production		%	11	14
Forestry losses	Production	m³	305,032	319,530
	Production (1.3 t/m³)	t	396,542	415,389
	Energy content (15 GJ/t)	PJ	5.94	6.23
Total biomass energy production		%	6	11

Table 9.4 Estimated crop energy content, crop residue factors and residue energy potential

Crop category	Crop energy content (GJ/t)	Waste factor	Residue energy content (GJ/t)
Fruits	3.2	2.0	13.1
Vegetables	3.2	1.0	6.0
Maize	14.7	1.4	13.0
Groundnuts	25.0	2.1	16.0
Coconuts	20.0	0.3	13.0
Roots and tubers	5.5	0.4	5.5
Sorghum	14.7	1.4	13.0
Sugar cane	5.3	1.6	7.7
Cocoa, beans	16.0	1.0	15.4
Coffee, green	7.6	1.0	14.2
Ginger	3.2	0.4	6.0

Waste factor and energy content values were obtained from Hemstock and Hall (1995, 1997), Amoo-Gottfried and Hall (1999) and Rosillo-Calle et al. (2006b). Roots and tubers consist of potatoes, sweet potatoes, cassava and yams.

content per tonne. However, the flow charts and Table 9.5 indicate that sugar cane is Fiji's highest producing crops, yielding a theoretical production of 14.7 times more than coconut (annual average 1999–2003) and 6.9 time more than coconut (annual average 2008–2012). The sugar cane residue 'bagasse' provides the highest amount of biomass energy used directly as fuel. Bagasse is used in

Table 9.5 Crop production, crop energy and residue energy for an annual average of 5 years

Crop category	Annual average crop production (tonnes)		Crop energy (GJ)		Annual average residue (tonnes)		Residue energy (GJ)		Total energy content (crop + residue) (GJ)	
	1999–2003	2008–2012	1999–2003	2008–2012	1999–2003	2008–2012	1999–2003	2008–2012	1999–2003	2008–2012
Fruits	35,761	37,992	114,436	121,576	71,522	75,985	936,943	995,401	1,051,379	1,116,977
Vegetables	16,889	18,399	54,044	58,877	16,889	18,399	101,333	110,395	155,377	169,273
Maize	720	815	10,581	11,978	1,008	1,141	13,100	14,829	23,681	26,807
Groundnuts	212	284	5,305	7,100	446	596	7,130	9,542	12,435	16,642
Coconuts	162,240	207,750	3,244,800	4,155,000	53,539	68,558	696,010	891,248	3,940,810	5,046,248
Roots and tubers	43,919	67,970	241,555	373,833	17,568	27,188	96,622	149,533	338,176	523,366
Sorghum	30	25	438	370	42	35	542	459	980	829
Sugar cane	3,291,000	1,979,400	17,442,300	10,490,820	5,265,600	3,167,040	40,545,120	24,386,208	57,987,420	34,877,028
Cocoa, beans	16	7	253	109	16	7	243	105	496	214
Coffee, green	14	16	108	125	14	16	202	233	310	358
Ginger	2,934	2,560	9,388	8,193	1,173	1,024	7,041	6,145	16,428	14,338

The annual average production was derived from FAOSTAT (2014); crop energy was derived using crop energy content (GJ/t) from Table 9.4; annual average residues were derived using the waste factor from Table 9.4; residue energy was derived using residue energy content in Table 9.4.

Table 9.6 Crop land use area in hectares

Crop category	Land area used (ha)	
	1999–2003	*2008–2012*
Fruits	2,404	3,634
Vegetables	2,148	1,737
Maize	345	402
Groundnuts	191	395
Coconuts	59,534	60,000
Roots and tubers	4,596	10,566
Sorghum	14	7
Sugar cane	64,000	46,600
Cocoa, beans	310	102
Coffee, green	27	30
Ginger	80	75

co-generation systems to provide power to Fiji's four sugar mills. Sugar cane residue has the highest source of energy, as shown in Figures 9.1 and 9.2. However, as is highlighted by the flow charts (Figures 9.1 and 9.2), the majority of the energy found in bagasse is not utilized and thus left in the field.

Livestock

The total potential energy from livestock were obtained by analysing the daily 'dung' production by respective stocks such as cattle, chickens, ducks, goats, horses, pigs, sheep and humans. The annual average production values were calculated using FAOSTAT-derived data as presented in Table 9.7.

Using the daily 'dung' production coefficient, an annual average for animal waste material (tonnes) was obtained for each type of livestock. Cattle have the highest 'dung' production capability. The annual 'dung' energy potential was obtained as: average annual 'dung' production (tonnes) × energy content (GJ/t). It was assumed that less than 1% of the 'dung' energy is used in biogas digesters as the factual data is not available to quantify the percentage of 'dung' energy utilized currently. It should be noted, however, that the Fiji Government (Departments of Agriculture and Energy) has an active project to promote and support farm-scale biogas digesters, and around 15 8 m³ Camatec design digesters have been built under this scheme.

What does the flow chart highlight?

The flow chart presented in Figure 9.1 for the years 1999–2003 shows that the total production of biomass energy was estimated at an annual average

Table 9.7 Livestock and human population, dung production and energy potential

FIJI stock	Average 5 year annual population (heads)	Dung production (dry weight kg/head/day)	Average annual dung (tonnes)	Energy content (GJ/t)	Annual dung energy potential (GJ)
1999–2003					
Cattle	327,000	1.80	214,839	18.5	3,974,522
Chickens	3,940,000	0.06	86,286	11.0	949,146
Ducks	74,000	0.06	1,621	11.0	17,827
Goats	236,770	0.40	34,568	14.0	483,958
Horses	43,940	3.00	48,114	11.0	529,257
Pigs	139,000	0.80	40,588	11.0	446,468
Sheeps	6,600	0.40	964	14.0	13,490
Humans	813,400	0.40	118,756	11.0	1,306,320
2008–2012					
Cattle	310,000	1.80	203,670	18.5	3,767,895
Chickens	4,640,000	0.06	101,616	11.0	1,117,776
Ducks	87,400	0.06	1,914	11.0	21,055
Goats	250,390	0.40	36,557	14.0	511,797
Horses	46,000	3.00	50,370	11.0	554,070
Pigs	145,620	0.80	42,521	11.0	467,731
Sheeps	6,200	0.40	905	14.0	12,673
Humans	860,000	0.40	125,560	11.0	1,381,160

Stock production figures were obtained from FAOSTAT (2014). Dung production and energy content were obtained from Amoo-Gottfried and Hall (1999), Rosillo-Calle et al. (2006b) and Mohammed et al. (2013).

of 86.17 PJ (74% from agricultural crop production, 17% from forestry and 9% from livestock). Of the 78.4 PJ produced from agricultural and forestry operations, 0.75 PJ of biomass was harvested and burnt as fuelwood, 21.2 PJ was harvested for food, 15.5 PJ was used to generate energy in co-generation (bagasse) and 32.7 PJ was un-utilized crop and forestry residues [forest residue does not consist of wood residue, which is principally an industrial residue, e.g. sawmill rejects, slabs, edgings and trimmings, veneer log cores, veneer rejects, sawdust, bark (excluding briquettes), residues from carpentry and joinery production, etc.].

The total amount of biomass (fuelwood, residues and 'dung') used directly to provide energy was estimated at 17.17 PJ (19.6 GJ per capita per year or 1.14 Mt fuelwood equivalent). Figure 9.1 accentuates biomass use and areas for potential use of biomass for energy and it is evident that losses from agriculture, forestry and livestock warrant further investigation as potential sources of biomass energy. However, it does not account for important issues such as collection efficiency, etc.

The flow chart presented in Figure 9.2 for the years 2008–2012 however showed a significant decline in biomass energy. The total production of biomass energy was estimated at an annual average of 65.30 PJ which has declined by 24%. Although production in wood cut, 'dung' and other agricultural crop indicated an increase, sugar cane production decreased significantly over the recent years as presented in Figure 9.3. Sugar cane biomass energy declined by 39% and this has a significant effect on the biomass energy potential in Fiji. The decline was a result of reduction in cultivation of sugar cane. The land area (in hectares) used for sugar cane farming declined by 27%, as can be seen in Table 9.6.

It is noted that farms which were previously occupied by farmers are now idle or are not used for full-capacity production. The reason for this is mainly due to land tenure issues specific to Fiji: native land leases were not renewed. To a lesser extent, some of the decline in production is because profits have decreased due to the high cost of inputs and harvesting/transport. However, looking at all agricultural crop production in Fiji, sugar cane still has the highest yield and area under crop so should be seen as the 'best option' when considering a source of biomass energy. Feedstocks such as cassava and sweet potatoes also have encouraging energy content; however, as shown in Table 9.8, both crops have a much lower production in terms of yield (tonne/ha) and thus lower energy potential (GJ/ha). A noted decrease in sugar cane production reveals that the industry has problems, such as decline in land used for production, that need to be addressed. Thus flow charts are an important 'broad-brush' tool to highlight the need for further investigation and highlight with the greatest potential for biomass energy production.

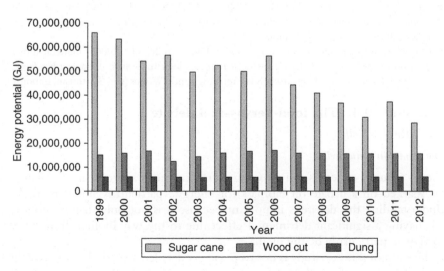

Figure 9.3 Production profile of sugar cane, wood cut and 'dung'.

Table 9.8 Annual crop production in Fiji, product energy content and residue energy content

Feedstock	Annual average production yield (tonne/ha)	Product energy content (GJ/tonne)	Product energy potential (gj/ha)	Residue energy content (GJ/tonne)	Residue energy potential (GJ/ha)	Total energy content (GJ/ha)
Cassava	12.84	5.6	71.90	13.1	67.28	139.18
Sugar cane	47.69	5.3	252.76	7.7	514.10	766.85
Sweet Potatoes	5.32	5.5	29.27	5.5	11.71	40.98

Annual average production yield was calculated using FAOSTAT for years 1999–2012. Product energy content for crops and residues were obtained from Hemstock and Hall (1995), Amoo-Gottfried and Hall (1999) and Rosillo-Calle et al. (2006a). Waste factor was obtained from table 9.4. Total energy content is a sum of product energy content and residue energy content.

References

AMOO-GOTTFRIED, K. AND HALL, D.O. (1999) A biomass energy flow chart for Sierra Leone. *Biomass and Bioenergy* 16, 361–376.

FAOSTAT (2014) http://faostat.fao.org/. Food and Agriculture Organization of the United Nations.

HEMSTOCK, S.L. AND HALL, D.O. (1995) Biomass energy flows in Zimbabwe. *Biomass and Bioenergy* 8, 151–173.

HEMSTOCK, S.L. AND HALL, D.O. (1997) A biomass energy flow chart for Zimbabwe: an example of the methodology. *Solar Energy* 59, 49–57.

MOHAMMED, Y.S., MOKHTAR, A.S., BASHIR, N. AND SAIDUR, R. (2013) An overview of agricultural biomass for decentralized rural energy in Ghana. *Renewable and Sustainable Energy Reviews* 20, 15–22.

ROSILLO-CALLE, F., DE GROOT, P. AND HEMSTOCK, S.L. (2006a) Appendices. In *Biomass assessment handbook*, 1st edn, F. Rosillo-Calle, P. de Groot, S.L. Hemstock and J. Woods (eds). Earthscan, London.

ROSILLO-CALLE, F., DE GROOT, P. AND HEMSTOCK, S.L. (2006b) General introduction to the basis of biomass assessment methodology. In *Biomass assessment handbook*, 1st edn, F. Rosillo-Calle, P. de Groot, S.L. Hemstock and J. Woods (eds). Earthscan, London.

Case study 9.4   The food-versus-fuel debate

Frank Rosillo-Calle

The debate in a nutshell

Energy security concerns and growing commitment to address climate change has sparked-off a debate on biomass for energy[1], particularly liquid biofuels. Understanding this debate is important because it is not only shaping policy, it is having a significant detrimental effect due to the way biofuels have been portrayed, particularly by the media.

The aim of this case study is not to present a detailed analysis, but to provide the key facts to inform the reader of the potential implications: the pros and cons of the debate. This case study is not populated with references; further information can be found in the Further reading section (from which most of the material originates) and elsewhere in this handbook.

The food–versus–fuel debate is not new, in fact it goes back to the 1970s. The overall debate has global origins and implications but there are three main geographical connotations: Brazil, EU and USA, with a disproportional emphasis on the USA. Further, the debate has had a narrow focus based on a handful of feedstocks, e.g. maize, cereals and sugar cane.

The biofuel debate has focused primarily on three main areas: (1) food versus biofuel production, (2) their positive and negative effects (i.e. GHG, climate warming and the broader environment) and (3) a socio-economic component. More recently the debate has also focused on the rapid expansion of solid biomass for electricity and heat generation, and potential adverse impacts of climate change.

The debate on biofuels has been particularly controversial because it has largely been driven by politics, ethical/moral considerations and vested interests rather than by science. Biomass for energy, particularly biofuels, has been plagued by unprecedented scrutiny while largely ignoring the consequences of other energy sources particularly fossil fuels; it is essential to have a fair playing field for all energy sources.

One of the stormiest issues is land use, direct and indirect (see Chapter 7). Land area required and the land category of using biomass for energy is at the heart of the debate. Indirect land use change (iLUC) has added uncertainty to benefits and impacts of bioenergy and biofuels because the methodology is still in its infancy and its application can lead to incorrect policy decisions with serious consequences for this sector.

This case study pays particular attention to five key areas: food prices, land use, energy balance, GHG and subsidies. It is argued that the use of biomass for energy on a large scale must go hand in hand with a modern, diversified, dynamic and scientifically driven agricultural sector. Superimposing a biofuel industry, for example, without fundamental changes in agriculture could spell disaster as it will be impossible to provide the huge additional demand on agriculture efficiently and sustainably.

Biomass for energy is both part of the problem and part of the solution. We need a holistic approach; a new paradigm is required to take full account of its diverse and complex nature, given its technical, economic, social and political ramifications, particularly for agriculture. Food security and biomass for energy can be mutually complementary rather than exclusive if the right policies are in place.

Much of the criticism focuses on the 'first-generation' (G1) technologies that probably will continue to prevail for at least the next decade until second-generation technologies (G2) become more widespread. There has also been a failure to fully understand the potential economic implications that additional demand for biofuel feedstocks places on agriculture commodity prices and human well-being. However, biofuels can bring considerable benefits if they meet the basic conditions of low GHG emissions from the whole supply chain and no, or very limited competition, with food crops.

The food–versus–fuel debate emanated first in the USA and in the EU. The use of corn in the USA in particular has been the subject of a vigorous debate for a combination of reasons, such as the existence of strong pro- and anti-biofuels lobbies, large-scale use of corn (a very poor ethanol feedstock), the powerful farmers' lobby and vested interest groups. This debate is embroiled in politics and hardy driven by science. The intensity of this controversy in both the EU and USA can be partly explained by the existence of specific policies put in place to subsidise large-scale biomass for energy, particularly biofuels.

The EU has been at the forefront of biomass for energy development and has been playing a key role in ensuring bioenergy/biofuels sustainability. The European Commission is acutely aware of social, environmental and policy implications if biomass for energy was to be imported in large scale. This leading role may be seriously undermined by constant policy shifts. There has been strong criticism both by the pro- and anti-biomass for energy lobbies. It is unclear what this potential policy change could mean but it could be serious. The food–versus–fuel debate has already resulted in a cap for conventional biofuels from 10 to 7% by 2020.

There are, basically two main schools of thought in this embracing debate: (1) the anti-biofuels lobby and (2) the pro-biofuels lobby. In a synopsis, the anti-biofuels lobby argument goes as follows.

- It is, simply, morally wrong to use land to produce biofuels instead of food when there are so many starving or undernourished people.
- Large-scale production of biofuels will lead to food insecurity worldwide.
- Increasing food prices will disproportionately affect the poorest people in developing countries, already overwhelmed with high food costs.
- Land competition, including iLUC, will increase with demand for food and non-food products, leading to deforestation, ecosystem destruction, loss of biodiversity and erosion of social rights.
- Many of the social and environmental benefits of biofuels are not yet fully proven.
- Large-scale production of biofuels will increase soil erosion, decrease carbon stocks, put stress on water resources and on other ecosystem services.
- Biofuels offer poor GHG benefits.
- Biofuels, in general, have a poor or negative energy balance.

Contrary, the pro-biofuels lobby arguments posits the following arguments.

- There is sufficient land available to produce both food and a reasonable portion of biofuels (i.e. 5–20% of transport fuels demand) without affecting food supply, with good and modern agricultural management practices.
- Large-scale biofuels production at a global level is highly unlikely; the focus will be in the most favourable geographical areas; physical availability of land *per se* is meaningless as there are many other non-physical determining issues, e.g. social inequality, productivity, policy, etc.

- Food-insecure countries that do not have their own fossil fuel reserves allocate a significant portion of their national income to pay for oil imports. In such cases, biofuels are a good alternative to fossil fuels and would free up foreign exchange for other investments.
- Nearly 2,000 million people live without access to modern energy systems. Agriculture in developing countries needs more energy in absolute terms and bioenergy availability can actually enhance food production.
- Multi-functional agricultural production systems already exist in most countries, are mutually beneficial and can produce both food and non-food products in a sustainable and socially balanced manner.
- Certain biofuels crops can be utilized to prevent soil degradation and can utilize or reclaim so-called 'marginal' or 'degraded' land.
- The social, economic, and environmental benefits of biofuels outweigh potential negative impacts if good management practices are applied.
- Bioenergy can spark off new investment in agriculture, increasing modernization and diversification and consequently productivity.
- Crop yields are, in general, very low and can be greatly enhanced.
- Biomass for energy can therefore play a significant role in modernizing and diversifying agriculture – e.g. as in the case of Brazil – and provide sustainable renewable energy.

The role of heat and power in the debate

Contrary to liquid biofuels, the concern with solid biomass is more recent and has its origins in the growing demand for heat and power generation in Europe but also in South East Asia, such as Japan and South Korea. This is because solid biomass is far more widely available and can use a wide range of feedstocks including, potentially, a large amount of residues and wastes that are currently mostly wasted. For example, a conservative global potential of residues has been put at between 300 and 18,000 TWh.

The key markets at present are primarily co-firing with coal in power plants (potentially the largest market) and the development of heating systems (all types). There is a large body of literature for the reader to explore further (e.g. see www.bioenergytrade.org/).

The key concerns with this market are the implications for the availability of sustainable feedstocks in large quantities if a large-scale global market develops (see Chapter 6), difficulties with logistics and the support policies (although these differ significantly depending of the country or region). This rapidly expanding market can have potentially large ramifications as hundreds of million tonnes of biomass could be traded internationally for energy (see also Case study 9.1).

The impact on food prices

The potential impacts of biofuels in food prices came dramatically to light in 2007–2008. This has been further ignited by the recent US drought, the worst in 50 years. Food prices are not static but fluctuate constantly in response to supply and demand. But the fact that the rapid development of biofuels has coincided with increases in food prices has not helped the industry. Similarly, the evolution of crude oil prices is also particularly striking. Thus, food price increases are the result of a complex web of factors that need to be incorporated into any debate. Here we mention just a few reasons, as follows.

- Changing consumption patterns leading to higher demand. Improved living conditions, particularly in developing countries such as Brazil, China and India, have led to an increase in demand for meat as consumers with growing wealth move away from eating just cereals.
- Distorted agricultural markets are caused by rich countries subsidising domestic agricultural production and off-loading surplus production on the world market, combined with low levels of investment in agriculture. This has been exacerbated in the last decade by low commodity prices as farmers struggle to survive.
- Increases in the price of crude oil, from US$50 to over $140 per barrel, have had a major impact on all chains of food production and distribution. There have been increases in the cost of inputs such as fertilizers and pesticides (linked to higher oil prices). Current low oil prices will be, most probably, short-lived.
- Rapid increases in feedstock demand from the biofuels sector were also a significant contributing factor, together with speculation in agricultural commodities by the financial sector.

Competition for land

The area and the category of land needed for growing biofuels have also been at the heart of the debate (see Chapter 7). The estimates of land use for biofuels, current and future, vary widely depending on the source and assumptions, multiple demands on land, etc., with estimates ranging from 40 to 800 Mha (currently there are about 38 Mha dedicated to biofuels). What seems certain is that estimates of future demand of land for biofuels production remain highly uncertain. It seems therefore that it is almost impossible to provide any reliable estimates at this point.

There is, however, some consensus that the potential for improving crop productivity is large; hence, to base potential production on the existing amount of land, without taking fully into account all the intertwined factors, is often meaningless. A major component of the land use debate has been iLUC impacts: this has added uncertainty for the biofuels industry. Although iLUC offers the prospect of reducing the negative effects, it must be applied equally to all products and fuels for there to be a level playing field. A large number of studies have investigated the impacts of iLUC on biomass for energy expansion, but with large discrepancies. The methodology for assessing iLUC needs to be improved significantly.

The key question is how to harness this potential economically and environmentally without compromising food supply. FAO data shows that, by 2050, 72 Mha of additional land will be needed to meet additional food demand, of which more than 40% will be for livestock feed.

GHG/environmental impacts

One of the greatest challenges currently facing humankind is the potential impact of climate change. Biomass for energy has widely been promoted as a partial solution to GHG problems. However, what role can biofuels play in limiting the level of GHG emissions? The answer depends upon the specific feedstock (e.g. sugar cane, maize, vegetable oil or lingo-cellulosic crops) and the circumstances of production and processing. Each crop has its pros and cons that need to be considered.

For ethanol, the highest GHG savings are recorded for sugar cane (70% to more than 100%), whereas corn can save up to 60% but may also cause 5% more GHG emissions, although there remain strong disagreements.

The conclusion of most of the scientific literature published in the recent past indicates that, on a Life-Cycle Analysis basis, biofuels provide a net GHG benefit (30–100% compared to petroleum fuels) when use of co-products is included and GHG emissions from land conversion are excluded in the analysis. These findings are, however, constantly being challenged as new data comes into light.

Energy balance

The energy balance expresses the ratio of energy contained in a fuel relative to the energy used in its production. A fossil energy balance of 1.0 means it requires as much energy to produce a litre of fuel as it contains. An energy balance of 2.0 means a litre of fuel contains twice the amount of energy as that required to produce it. Balances for biodiesel range from around 1 to 4 for rapeseed and soybean feedstocks, whereas palm oil is around 9. For crop-based ethanol, balances range from less than 2 for maize to around 2–9 for sugar cane (8–10 in the case of Brazil). Sugar-cane-based ethanol, as produced in Brazil, depends not only on feedstock productivity but on the fact that it is processed using biomass residues (sugar cane bagasse) as an energy input. The energy balances for cellulosic feedstocks tend to be even higher.

The energy balance of biofuels production is nevertheless still a contentious issue, particularly for ethanol from corn. It is often oversimplified, given the complex web of economic, technical and political factors that need to be taken fully into consideration; different assumptions and calculations can, therefore, lead to different results. Major improvements (i.e. greater crop productivity, reduced energy consumption, greater use of bioenergy, development of new co-products, etc.) will further improve the energy balance.

Bioenergy subsidies

Subsidies given to biomass for energy have been sharply criticized but need to be examined in a wider context. Government subsidies have supported energy production, both for fossil fuels and renewable energy, for many decades. The size of these subsidies varies considerably from country to country. Historically subsidies given to fossil fuels and nuclear energy have been orders of magnitude greater than for biofuels. The main problem with current fossil fuel subsidies is that their hidden environmental costs are not often internalized and neither are their potential climate impacts.

The International Energy Agency (IEA) has conducted many studies on the effects of energy subsidies and the benefits of removing them. For example, in 2008 the amount of subsidies given to fossil fuels totalled US$557 billion, representing 2.1% of these countries' GDP ($312 billon for oil, $204 billion for natural gas and $40 billion for coal).

Critics suggest that the subsidies to biofuels, for example, distort the market while ignoring large subsidies given to the energy sector in general and oil in particular. Oil, primarily diesel and petrol, are still heavily subsidised in many developing countries, especially oil-producing ones. What is needed is to have a fair playing field for all energy sources, not just for biomass for energy.

Potential climate-induced changes

Concern is growing about a possible global food crisis, largely based on potential adverse climatic conditions, e.g. recent droughts that ravaged crops in the USA (and also floods) in Australia, India, Russia, Ukraine, etc. In this respect, a huge political debate has started in the USA about the future of ethanol. A large group of US law-makers called to halt or lower the subsidy ruled on how much ethanol the country could use. Although ethanol retains strong bipartisan political support, this impacted negatively in the USA and beyond. Also, various United Nations agencies (the Food and Agriculture Organization, World Bank and other organizations) have added their voices to suspending ethanol quotas as a response to the impact of the worst US drought on corn supplies and prices.

The recent drought and the record high grain prices have also strengthened considerably the anti-biofuel lobby in the EU. For example, the EU's Executive Commission is under considerable pressure to forge a deal to help ensure that EU biofuels do not clash with food production or the environment, calling for a more flexible policy. In the EU far more than in the USA, the *raison d'être* of biofuel is to lower carbon emissions.

Tackling agriculture and huge food waste

Many studies, particularly the anti-biofuels lobby, tend to overlook that the present agricultural production, food and feed, transportation, etc., is hugely inefficient and terribly wasteful. For example, it is estimated that 30–50% of global food produced rots away, uneaten. In some countries as much as 75%

of the harvest is lost, particularly fresh vegetables. A recent report by the US Natural Research Defence Council says Americans throw away nearly half their food every year, wasting roughly $165 billion annually! And what is particularly worrisome is that there has been a 50% jump in US food waste since the 1970s. Tackling this huge waste must be at the core of any government's policy.

Main conclusions

There are no simple answers to potential problems posed by biomass for energy; it will be very difficult to avoid conflict for food, feed and fuel because of the increasing demand for agricultural-based products, changing diets, limited scope for rapid increase in yields, etc. This requires bold new initiatives of how we produce and consume food and fuels.

In the short term, G1 biofuels represent the best and most realistic partial alternative to oil as transport fuel; in the medium to long term there exist various other alternatives offered by the G2 and G3, in the form of biofuels, bioelectricity, hydrogen, etc. The same applies to bioenergy used to generate power and heat when substituting coal. Despite some negative effects of biomass energy it is well established that, on balance, the positive factors far outweigh the negative ones.

Biomass energy is both part of the problem and part of the solution. A new paradigm is required to take full account of its complexity given the technical, economic, social and political ramifications, particularly for agriculture. It is important to move beyond the current debate to incorporate complementarity and modern agricultural management, and huge food waste. Such debate must be based fundamentally on scientific facts as much as possible, away from ethical and political considerations. Biomass can make a large contribution to the world's future energy requirements; this is a resource we cannot ignore. The challenge is to harness it in an environmental and economical manner and without compromising food security.

Note

1 In this Appendix, biomass for energy denotes bioenergy and biofuels; bioenergy refers to solid biomass used primarily for domestic uses (heating, cooking) and industrial applications (heat and power), for both small and large scale uses; biofuels refer to liquid biofuels (biodiesel and bioethanol) used primarily in road transport.

Further reading

ROSILLO-CALLE, F. (2010) Food versus fuel: can we avoid conflict? In *Sugarcane bioethanol: R&D for productivity and sustainability*, L.B. Cortez (ed.). Blucher, Sao Paulo, pp. 101–113.

ROSILLO-CALLE, F. (2012) Food versus fuel: toward a new paradigm: the need for a holistic approach. *ISRN Renewable Energy* 2012, article ID 954180.

ROSILLO-CALLE, F. AND JOHNSON, F. (eds) (2010) *The food versus fuel debate: an informed introduction to biofuels*. Zed Books, London.

10 General technical appendices

Appendix 10.1 Glossary of related terms

Accessible fuelwood (woodfuel) supplies The quantities of wood that can actually be used for energy purposes under normal conditions of supply and demand.

Advanced biofuels Refers to advances in crop and conversion technologies which enable higher yields from non-food crops and production of biofuels with better performance and that are more environmental friendly than traditional biofuels.

Agrofuels Are biofuels obtained either as a product or by-product of agriculture; it covers mainly biomass materials derived directly from *fuel crops* and *agricultural, agro-industrial and animal by-products* (FAO).

Air-dried weight This is the weight of wood in an air-dry state, after being exposed over time to local atmospheric conditions. Weight of wood can be given in an air-dry state or wet state. It may contains between 8 and 12% (dry basis) moisture.

Air-dried yield (wood) The approximate mass of air-dry wood which would be obtained upon drying or wetting of a sample of wood per unit volume of a sample.

Air-dry The term refers to the stage after the fuel has been exposed for some time to local atmospheric conditions, between harvesting and conversion of the fuel either to another fuel or by combustion to heat energy.

Air-dry density The density based on the weight and volume of wood in equilibrium with the atmospheric conditions.

Alcohol fuels A general term which denotes mainly ethanol, methanol and butanol, usually obtained by fermentation, when used as a fuel.

Ancillary energy The energy required to produce inputs to agricultural production consumed in a single activity or production period; e.g. the energy sequestered in fertilizers, chemicals, etc.

Animal waste The dung, faeces, slurry or manure which is used as the raw material for a biogas digester.

Animate energy Work performed by animals and humans. This is a very important source of power in many developing countries for agriculture and small-scale industry. Animals and humans provide pack and draught power, and transport by bicycle, boat, etc. (See Appendix 4.3, Measuring animal draught power).

Ash content The weight of ash expressed as a percentage of the weight before burning of a fuel sample under standard conditions in a laboratory furnace. The higher the ash content, the lower the energy value of the fuel.

Bagasse The fibrous residue from sugar cane which remains after the juice has been extracted. It constitutes about 50% of cane stalk by weight and with a moisture content of 50% its calorific value varies from 6.4 to 8.60 GJ/t. It widely used to generate electricity, and also as animal feed, ethanol production, pulp and paper, paperboard, furniture, etc.

Bark A general term for all the tissues outside of the cambium in stems of trees; the outer part may be dead, the inner part is living.

Barrel A measurement used in the oil industry that equals 159 litres (42 US gallons).

Basal area The cross-sectional area of a tree estimated at breast height; it is normally expressed in square metres. The sum of the basal areas of trees on an area of 1 ha is symbolized by $G = m^2/ha$. (Basal area is usually measured over bark.) Common values in young plantations are 10–20 m^2 rising to a maximum of around 60 m^2/ha in exceptional circumstances in older plantations.

Biobutanol An advanced biofuel that offers some potential benefits over conventional bioethanol and other conventional fuels, with an energy density close to petroleum. It also offers the advantage of high blends with conventional fuels.

Bioenergy, biomass energy Covers all energy forms derived from organic fuels (biofuels) of biological origin used for energy production. It comprises purpose-grown energy crops as well as multi-purpose plantations and by-products (residues and wastes). The term by-products covers solid, liquid and gaseous by-products derived from human activities. Biomass may be considered as one form of transformed solar energy (FAO). There are two main types of biomass energy: **biomass energy potential** and **biomass energy supply**. See under these terms. See also **Biomass for energy**.

Bioethanol See **Ethanol fuel**.

Biofuel A general term which includes any solid, liquid or gaseous fuel produced from organic matter, either directly from plants or indirectly from industrial, commercial, domestic or agricultural wastes.

Biogas The fuel produced following the microbial decomposition of organic matter in the absence of oxygen. It consists of a gaseous mixture of methane and carbon dioxide in an approximate volumetric ratio of 2:1. In this state the biogas has a calorific value of about 20–25 MJ/m^3 but this can be upgraded by removing the carbon dioxide.

Biomass The organic material derived from biological systems. Biological solar energy conversion via the process of photosynthesis produces energy in the form of plant biomass which is about 10 times the world's annual use of energy. Biomass does not include fossil fuels, although they also originate from biomass-based sources. For convenience, biomass might be sub-divided into two main categories: **woody biomass** and **non-woody biomass** (see under these terms) although there is no clear distinction between these terms.

Biomass conversion process   The methods which convert biomass into fuel can be classified as: (1) biochemical, which includes fermentation and anaerobic digestion; (2) thermo-chemical, which includes pyrolysis, gasification and liquefaction. See under these names.

Biomass energy potential   This term refers to the total biomass energy generated per annum. This represents all the energy from crop residues, animal wastes, harvestable fuel crops and the annual increase in the volume of wood in forests.

Biomass energy supply   The term represents the total biomass that is accessible to the market taking into account the logistics of collecting from various sources.

Biomass energy trade or **biotrade**   A general term denoting the commercialization of all biomass-based fuels, traded either locally, nationally or internationally.

Biomass for energy   This denotes both bioenergy and biofuels; bioenergy refers to solid biomass used primarily for domestic uses (heating, cooking) and industrial applications (heat and power), for both small- and large-scale uses; biofuels refer to liquid biofuels (biodiesel and bioethanol) used primarily in road transport.

Biomass inventory   All living organisms and dead organic material in the soil (humus) and in the sea. Ninety-nine per cent of living organic material is plant biomass, produced mainly by forests, woodland, savannahs and grassland.

Biomass productivity   The increase in wet or dry weight of living and dead plant material/unit area or individual/unit time; biomass production may be expressed, e.g. on the basis of the whole tree or parts thereof.

Biomass sustainability   The UN World Commission on Environment and Development (UN-WCED) *Our Common Future* (1987) document defined sustainable development as 'Development that meets the needs of the present without compromising the ability of future generations to meet their own needs'. The same principle applies to biomass sustainability, although this term can be broken into different components, e.g. environmental, ecological, economic and social. Sustainable development of biomass energy resources has become a fundamental determining factor.

Bole   Primary length of a tree, the trunk of a tree.

Bone dry   See **oven-dried**.

Breast height diameter   See **Diameter at breast height**.

Brix   Scale of densities used in the sugar industry. Hydrometers are marked in 'degrees Brix', representing the density of a corresponding pure sugar solution in unit equivalent to the percentage of sugar in the solution, either by volume ('volume Brix') or by weight (weight Brix). World average is around 15 or 16. See also **pol**.

Brushwood　Shrubby vegetation and stands of tree species that do not produce commercial timber.

Bulk density　The mass (weight) of a material divided by the actual volume it displaces as a whole substance expressed in kg/m^3, etc.

Burning index　A number on an arithmetical scale, determined from fuel moisture content, wind speed and other selected factors that affect burning conditions, and from which ease of ignition of fires and their probable behaviour may be estimated.

Calorific value　A measure of the energy content of a substance determined by the quantity of the heat given off when a unit weight of the substance is completely burned. It can be measured in calories or joules; the calorific value is normally expressed as kcal/kg or MJ/kg.

Canopy　The total leaf cover produced by branches of forest trees and leaves of other plants, affording a cover of foliage over the soil.

Charcoal　The residue of solid non-agglomerating organic matter, of vegetable or animal origin, that results from carbonization by heat in the absence of air at a temperature above 300°C.

Combustion energy　The energy released by combustion. It is normally taken to mean the energy released when substances react with oxygen. This may be very fast, as in burning, or slow, as in aerobic digestion. Technical temperature achieved depends on whether combustion is with air or pure oxygen, and on the method of combustion.

Commercial forestry　The practice of forestry with the object of producing timber and other forest produce as a business enterprise.

Conventional energy (or **fuel**)　Denotes energy sources which have hitherto provided the bulk for the requirement of modern industrial society, e.g. petroleum, coal and natural gas; wood is excluded from this category. The term is almost synonymous with *commercial energy*.

Conversion efficiency　The percentage of total thermal energy that is actually converted into electricity by an electricity-generating plant.

Cookstove　Used widely for cooking food in developing nations. There are many types; they use mostly biomass fuels, particularly wood. Stoves have very low thermal efficiency, about 13–18%.

Cord　A measure of stacked wood. By definition 1 cord is a stack of pieces of wood which is 4 feet wide, 4 feet high and 8 feet long, giving a total of 128 cu. ft. In practice the weight and volume in a cord can vary appreciably.

Crop residue index (CRI)　A method used for estimating crop residues. This is defined as the ratio of the residue produced to the total primary crop produced for a particular species. The biomass produced by crop plants is usually one to three times the weight of the actual crop itself. The CRI is determined in the field for each crop and crop variety, and for each agro-ecological region under consideration.

Current annual increment (CAI)   The increment of total biomass produced over a period of 1 year. It must be distinguished between increment of the standing crop and net production. The difference is the amount loss by litter fall, root slouching and grazing.

DBH   See **diameter at breast height**.

Decentralized energy   The energy supplies generated in dispersed locations and used locally, maintaining a low energy flux from generation to supply. The term is often used for renewable energy supplies since these harness the energy flows of the natural environment, which are predominantly dispersed and have relatively low energy flux. In contrast, the centralized energy supplies of large-scale fossil and nuclear sources produce large energy fluxes which are most economically used in a concentrated manner.

Delivered energy   The actual amount of energy available or consumed at point of use. The concept recognizes that in order to have a unit of economically usable energy, there are prior exploration, production and delivery systems, each of which detracts from the next amount of energy delivered to the point of use. It is also called *received energy* since it records the energy delivered to or received by the final consumer.

Densified biomass fuels   Fuel made by compressing biomass to increase the density and to form the fuel into a specific shape such as briquettes, pellets, etc. to ease transportation and burning.

Density   The weight of unit volume of a substance. In the case of wood several different densities can be referred to: (1) basic density (the weight of dry matter in unit volume of freshly felled wood); (2) air-dry density (the weight of unit volume of oven-dry wood); (3) stacked density (the weight of wood at stated moisture content – fresh, air dry, etc. – contained in a stack of unit volume).

Diameter at breast height (DBH)   Method used by foresters to determine total height and crown measurements (diameter + depth) to estimate individual tree volumes, mean height, basal area at breast height and mean crown measurements. This allows the tree crop volume per unit area to be estimated. It is the reference diameter of the main stem of standing trees, usually measured at 1.3 m above the ground level. It is one of the measurable dimensions from which tree cross-sectional area and volume can be computed.

Direct combustion   Complete thermal breakdown of organic material in the presence of air so that all its energy content is released as heat.

Direct land use change (dLUC)   Occurs when crops for biofuel production are planted on land that has not previously been used for that purpose, for example, the conversion of forest into an energy crop plantation for biofuels production. The effects of dLUC can be directly observed and measured as the effects are localized to a specific plantation. See also **Land use management**.

Dry basis   A basis for calculating and reporting the analysis of a fuel after the moisture content has been subtracted from the total. For example, if a fuel sample contains A% ash and M% moisture, the ash content on a dry basis is: 100:100 − A%.

Dry fuel   Biomass material with a low moisture content, generally 8–10%. The allowable moisture content for dry fuel varies with requirements of the combustion system.

Dry tonne   Biomass dried to a constant weight of 2000 lb.

Energy content   The intrinsic energy of a substance, whether gas, liquid or solid, in an environment of a given pressure and temperature with respect to a data set of conditions. Any change of the environment can create a change of the state of the substance with a resulting change in the energy content. Such a concept is essential for the purpose of calculations involving use of heat to do work. See also **heating value**.

Energy content as received   The energy content of a fuel just before combustion. It reflects moisture content losses due to air-drying or processing. For this reason, the energy content as received is generally higher per unit weight than of the fuel at harvest.

Energy content of fuel at harvest   Normally used for biomass resources, it refers to the energy content of a fuel at the time of harvest. It is also referred to as the *green* energy content.

Energy efficiency   The percentage of the total energy input that does useful work and is not converted into low-quality, essentially useless, low-temperature heat in an energy conversion or process.

Energy grasses   Perennial high-fibre grasses specifically grown for energy purposes, such as *Miscanthus*.

Ethanol fuel (bioethanol)   Fermentation ethanol obtained from biomass-derived sources, usually sugar cane, maize, etc. used as a fuel. Ethanol is also obtained from fossil fuel, e.g. coal and natural gas.

Fermentable sugar   Sugar (usually glucose) derived from starch and cellulose that can be converted to ethanol. Also known as reducing sugar or monosaccharide.

Firewood   See **woodfuel**.

Food versus fuel   A term embracing a general debate concerning the implications of the utilization of land to produce fuel rather than food. The debate has embraced considerable moral/ethical and political issues rather than fundamental scientific facts.

Forest inventory   Forests have different structures; that is, the different species, ages and sites of trees are grouped in different patterns in different forests. An inventory of a forest area can provide information for many different purposes, e.g. it may be part of a natural resource survey, a national project to assess the potential of forest, etc. Forest inventory thus refers techniques used for measuring of trees.

Forest mensuration   Branch of forestry concerned with the determination of the dimensions, form, increment and age of trees, individually or collectively, and of the dimensions of their products, particularly sawn timber and logs.

Fuelwood (or **firewood**)   See **woodfuel**. Fuelwood is the term used by FAO.

Fuelwood needs   Minimum amount of fuelwood/woodfuel necessary in view of the minimum energy estimated to be indispensable for household consumption, artisanal purposes and rural industries, in line with local conditions and the share of fuelwood in their energy supplies.

Fuelwood supply   Is the quantity of fuelwood/woodfuel which is, or can be made available for energy use on the basis of the mean annual productivity of all potential resources on a sustainable basis.

Green fuel   A term that denotes freshly harvested biomass not substantially dehydrated, with varying moisture content, obtained from processed vegetal wastes and organic material. It also denotes biomass-based fuels.

Green manure   Fresh or still-growing green vegetation ploughed into the soil to increase the organic matter and humus available to support crop growth.

Green weight   The weight of freshly cut wood, which contains 30–35% (dry basis) moisture.

Gross heating value (GHV)   It refers to the total heat energy content of a fuel; it equals the heat released by complete combustion under conditions of constant volume. It is sometimes referred to, erroneously, as the *higher heating value*.

Gross increment   Average volume of increment over given period of all trees (all diameters down to a stated minimum diameter). Also included is the recruitment (in growth) of small trees when they reach the minimum diameter.

Gross primary production (Pg) (biomass)   The total amount of photosynthetically produced organic matter assimilated by a community or species per unit land area per unit time.

Growing stock   Refers to the living part of the standing volume of trees (see **standing volume**).

Harvest index   The index of a given crop is obtained by dividing the weight of the useful harvested part by the total weight of organic material produced by the crop.

Heating value   The heating value (HV) of fuels is recorded using two different types of energy content: (1) **gross heating value (GHV)** and (2) **net heating value (NHV)**. See under these entries. Although for petroleum the difference between the two is rarely more than 10%, for biomass fuels with widely varying moisture contents the difference can be quite large. Heating values of biomass fuels are often given as an energy content per unit weight or volume at various stages: **green fuel**, **air-dried** and **oven-dried** material. See under these names. See also **energy content**.

Indirect land use change (iLUC)   It is considered to occur when, as a result of switching agricultural land use to biofuel crops, a compensating land use change occurs elsewhere to maintain the previous level of agricultural production. These effects are typically the unintended consequence of land use decisions elsewhere and, given that the effects are not limited by geographical boundaries (e.g. the complex dynamics of food commodities worldwide), often are not directly observable or measurable. See also **land use management**.

Land use management The process of assessment of land performance when used for specified purposes involving the execution and interpretation of surveys and studies on all aspects of land in order to identify and make a comparison of promising kinds of land use in terms applicable to the objectives of the evaluation.

Leaf area index (LAI) The leaf surface area (one side only) in a stand of a given crop covering a unit area of land surface. Optimum photosynthesis conditions are achieved.

Lower heating value (LHV) See **net heating value**.

Maximum sustainable yield The largest amount which can be taken from a renewable natural-resource stock per period of time, without reducing the stock of the resource.

Mean annual increment (MAI) Average annual increment over a given number of years; it is obtained by dividing the total biomass produced by the number of years taken to produce it.

Measurement units There are four basic types of unit used in energy measurements: (1) **stock energy units**, (2) **flow or rate energy units**, (3) **specific energy consumption or energy intensity** and (4) **energy content** or **heating value**. See further information under the bold terms.

Moisture The moisture content is the amount of water contained in a fuel. For biomass fuels special care must be taken to measure and record the moisture content. The moisture content can change by a factor of 4–5 between initial harvesting and final use and is critical both to the heating value on a weight or volume basis and to differences between GHV and NHV. The approximate moisture contents typical of various woodfuels are, e.g. 44–23 wet-weight and 80–30 dry basis for hardwood whole tree chips, green sawdust.

Moisture content dry basis (mcdb) The ratio of the weight of water in a fuel to the oven-dry (solid fuel) weight, expressed as a percentage.

Moisture content wet basis (mcwb) The ratio of the weight of water in a fuel to the total weight of the fuel. The mcwb is expressed as percentage.

Net energy ratio (NER) The NER of a given biomass fuel process may be calculated by dividing the heat content of the final fuel product by the energy (heat) equivalent of all the processes and machinery that were used in the process.

Net heating value (NHV) The potential energy available in the fuel as received, taking into account the energy that will be loss in evaporating and superheating the water when the woodfuel burns. This energy loss accounts for most of the reduction in efficiency when systems are fuelled with green wood rather than dry wood or fossil fuels. It is sometimes referred to as the *lower heating value*.

Net increment Average net increment less natural losses over a given period.

Net primary production (Pn)   The total amount of photosynthetically produced organic matter assimilated less that loss due to plant respiration, i.e. the total production which is available to other consumer: trophic levels or that which remain stored as chemical energy. Primary production is the production of vegetation and other autotrophic organisms. Pn denotes a gain of material by a plant community; it can be determined from the sum of changes in plant biomass together with losses of plant material, e.g. death, etc., over a given time interval. Pn is measured typically as grams per unit area or volume. See also **primary productivity**.

Net primary productivity   Refers to the amount of organic matter formed by photosynthesis per unit of time, i.e. the amount remaining in the plant after respiration. Usually expressed in terms of dry weight.

Non-woody biomass   There is no clear division between this term and **woody biomass** due to the nature of biomass sources. For example, cassava, cotton and coffee are all agricultural crops but their stems are wood. For convenience, non-woody biomass includes mostly agricultural crops, shrubs and herbaceous plants. See also **woody biomass**.

Normalized vegetation index (NVI)   Refers to the ratio of the red to near infrared reflection from vegetation as detected from remote sensing.

Oven-dried   Wood dried at constant weight in a ventilated oven at a temperature above the boiling point of water, generally 102–103°C. It means that the fuel has zero moisture content and is sometimes referred to as bone dry.

Oven-dried weight   The weight of a fuel or biomass material with zero moisture content.

Oven wood   The wood that has been split into fairly short, thin pieces and then air-dried.

Photosynthesis   A term used commonly to denote the process by which plants synthesize organic compounds from inorganic raw materials in the presence of sunlight. All forms of life in the universe require energy for growth and maintenance. It is therefore the process whereby green plants use the sun's energy to produce energy-rich compounds, which may then be used to fix carbon dioxide, nitrogen and sulphur for the synthesis of organic material.

Pol (short for polarization)   An analytical method to determine the level of sucrose in sugar cane. It is the apparent sucrose content of any substance expressed as a percentage by mass; it is often cited as a mixed juice pol, with world average value around 12 or 13. See also **Brix**.

Primary energy (or **primary production**)   It measures the potential energy content of the fuel at the time of initial harvest, production or discovery prior of any type of conversion. It is often used for recording the total energy consumption of a country, which is misleading because it ignores the conversion efficiencies at which the fuel is used.

Primary productivity   The next yield of dry plant matter produced by photosynthesis within a given area and period of time. Photosynthesis efficiency is the main determining factor in primary productivity. The annual global primary productivity is from $100-125 \times 10^9$ tonnes on land plus $44-55 \times 10^9$ tonnes in the world's oceans. Most of this biomass (44.3%) is formed in forests and woodlands.

Random sampling In a random sampling the choice of each sampling unit selected for measurement is made independently of that of many others; that is, the selection of any one unit gives no indication of the identity of any other selected unit. Selection must embody the Laws of Change so that each unit in the sampling frame has a known possibility of selection.

Remote sensing Surveying the Earth's surface from aircraft or satellites using instruments to record different parts of the electromagnetic spectrum; it is also used to measure total biomass productivity. There is more than one definition and an array of techniques (see Chapter 8).

Renewable energy Refers to an energy form the supply of which is partly or wholly generated in the course of the annual solar cycle. The term covers those continuous flows that occur naturally and repeatedly in the environment; e.g. energy from the sun, the wind, from plants, etc. Geothermal energy is also usually regarded as a renewable energy source since in total it is a resource on a vast scale.

Renewable resources Natural resources produced by photosynthesis, or derived from products of photosynthesis, e.g. energy from plants; or direct from the sun, e.g. solar energy utilized by humans in the form of plant or animal products. See **renewable energy**.

Secondary energy (or **secondary fuels** or **final energy**) It defers from **primary energy** by the amount of energy used and loss in supply-side conversion systems; that is, sources of energy manufactured from basic fuels, e.g. biomass gasifiers, charcoal kilns.

Site class A measure of the relative productive capacity of a site for the crop or stand under study based on volume or weight or the maximum mean annual increment that is attained or attainable at a given age.

Site index See **site class**.

Solid volume (wood) Volume of the actual logs only by taking individual dimensions of evenly cut geometries of these logs; usually measured in cubic metres.

Stacked volume (wood) Volume occupied by logs when they are stacked closely to form heaps of given dimensions. Usually measured in cubic metres.

Stand A continuous group of trees sufficiently uniform in species composition, arrangement of age classes, and conditions to be a homogeneous and distinguishable unit.

Standing volume The volume of standing trees, all species, living or dead, all diameters down to a stated minimum diameter. Species which do not have an upright trunk (brush, etc.) are not considered as trees. It includes dead trees lying on the ground which can still be used for fibre or as fuel.

Standing volume tables   Instead of felling sample trees or measuring them standing or using a single tree volume table, volume per hectare of even-aged crops of a single species may be predicted directly using a stand volume table. The commonest stand volume table is derived from simple linear regression of volume per hectare, V, or on the combined variable-basal area per hectare multiply by some measure of height representative of the crop; often dominant height is used because it is convenient, being objectively defined as the height of the 100 fattest or tallest trees per hectare. Standing volume tables normally have confidence limits at 0.05 of about ±5–10% of the mean stand volume using 20 samples per stand.

Stere   A measure of stacked wood. By definition a stere is a stack of wood 1 m long by 1 m wide by 1 m high having a total volume of 1 m^3. In practice the weight of a stere ranges between about 250 and 600 kg.

Stock   The total weight of biomass at any given time. Preferably, biomass stock or inventory should be expressed initially as total gross figures, and then as even dried weight.

Stock energy units   Measure a quantity of energy in a resource or stock, e.g. wood energy in a tree at a given point in time. Examples are: tonnes of oil equivalent or multiples of the joule (MJ, GJ, PJ).

Stocking (wood)   The volume or weight of woody material on unit area, often expressed as a proportion of a measured or theoretical maximum possible under the particular soil and climatic conditions prevailing. Often the proportion of the unit area covered by tree crowns is used as a rough guide to stocking.

Stumpage   The value of a standing tree. The terms *stumpage fee, stumpage value, royalty* and *stumpage tax* are often used interchangeably, often causing confusion. Stumpage fee is the financial concept of the value of standing wood resources. Stumpage value is the economic concept of the worth to society of a unit of wood resource, estimated using the economic concepts of real resource (opportunity) costs and shadow (efficiency) prices. Stumpage royalty (royalty), presently refers to payments extracted by a public authority (e.g. government) in exchange for use of a tree on public land.

Total biomass volume   All above-ground parts of a plant.

Tree height measurement   Tree height measurement is important as it is often one of the variables commonly used in the estimation of tree volume. Tree height can have different meanings which can lead to practical problem. Thus it is important that the terms are clearly defined. *Total height* refers to the distance along the axis of the tree stem between the ground and the tip of the tree. *Bole height* is the distance along the axis of the tree stem between the ground and the crown point. *Merchantable height* refers to the distance along the axis of the tree stem between the ground and the terminal position of last usable portion of the tree stem (minimum stem diameter). *Stump height* is the distance between the ground and the basal position on the main stem where a tree is cut, about 30 cm. Merchantable length refers to the distance along the axis of the tree stem between the top of the stump and the terminal position of the last usable portion of the tree stem. *Defective length* is the sum of the portions of the merchantable length that cannot be utilized due to defects. *Sound merchantable length* equals the merchantable length minus the defective length. *Crown length* refers to the distance on the axis of the tree stem between the crown point and the tip of the tree.

Vegetation indices These indices are derived from combined visible and near infrared observations from terrestrial remote sensing. Vegetation indices are an important aspect of electromagnetic remote sensing which may be of great value to bioclimatic research.

Woodfuels All types of biofuels derived directly and indirectly from trees and shrubs grown in forest and non-forest lands. Woodfuels also include biomass derived from silvicultural activities (thinning, pruning etc.) and harvesting and logging (tops, roots, branches, etc.), as well as industrial by-products derived from primary and secondary forest industries that are used as fuel. They also include woodfuel derived from *ad hoc* forest energy plantations. Woodfuels are composed of four main types of commodities: fuelwood (or firewood), charcoal, black liquor and other.

Woodfuels (direct, indirect and recovered) According to origin, wood-fuels can be divided into three groups: direct woodfuel, indirect woodfuel and recovered woodfuel. *Direct woodfuel* consists of wood directly removed from forests (natural forests and plantations; land with tree crown cover of more than 10% and area of more than 0.5 ha), other wooded lands (land either with a tree crown cover of 5–10% of trees able to reach a height of at least 5 m at maturity *in situ*, or crown cover of more than 10% of trees not able to reach a height of 5 m at maturity *in situ*, and shrub or bush cover) and other lands to supply energy demands and includes both inventoried (recorded in official statistics) and non-inventoried woodfuels. *Indirect woodfuels* usually consist of industrial by-products, derived from primary (sawmills, particle boards, pulp and paper mills) and secondary (joinery, carpentry) wood industries, such as sawmill rejects, slabs, edging and trimmings, sawdust, shavings and chips bark, black liquor, etc. *Recovered woodfuels* refer to woody biomass derived from all economic and social activities outside the forest sector, usually wastes from construction sites, demolition of buildings, pallets, wooden containers and boxes, etc. (FAO).

Woody biomass It comprises mainly trees and forest residues (excluding leaves), although some shrubs and bushes and agricultural crops such as cassava, cotton and coffee stems are wood and are often included. Woody biomass is the most important form of biomass energy. See also **biomass** and **non-woody biomass**.

Yield For plant matter yield is defined as the increase in biomass over a given time and for a specific area, and must include all biomass removed from the area. The yield or annual increment of biomass increment is expressed in dry tonnes/hectare per year. It also should be clearly stated whether the yield is the current or mean annual increment.

Appendix 10.2   Most commonly used biomass symbols

Biomass and other

od, OD	oven dry
odt, ODT	oven dry tonne
ad, AD	air dry
mc, MC	moisture content
mcwb	moisture content, wet basis
mcdb	moisture content, dry basis
MAI	mean annual increment
GHV	gross heating value
HHV	high heating value
LHV	lower heating value
NHV	net heating value

Metric unit prefixes

Prefix	Symbol	Factor
exa	E	10^{18}
peta	P	10^{15}
tera	T	10^{12}
giga	G	10^{9}
mega	M	10^{6}
kilo	k	10^{3}

Metric equivalents

1 km	1,000 m
1 m	100 cm
1 cm	10 mm
1 km²	100 ha
1 ha	10,000 m³

Costs

Multiply	by	To obtain
$/tonne	1.1023	$/Mg
$/Mg	0.9072	$/tonne
$/MBtu	0.9470	$/GJ
$GJ	1.0559	$/MBtu

Appendix 10.3 Energy units: basic definitions

Energy can be measured in different units, among which the most generally used is *joule* (symbol: J). In the past, *calorie* (symbol: cal) was commonly used to measure heat (thermal energy). However, the 9th General Conference on Weights and Measures in 1948 adopted the joule as the unit of measurement for heat energy.

The energy content of *commercial energy sources* is expressed by the *calorific values*, which give the energy content of the given material related to the unit of weight, measured in *kJ/kg* (formerly in kcal/kg). The calorific values of various energy sources are always approximate figures since their precise values greatly depend on their chemical composition.

The quantity of *liquid fuels* and some solid fuels, like fuelwood, wood chips, etc. in many cases are measured by their volume and consequently, their calorific values can be expressed in relation to their volumes as well. With the help of the *specific gravity* (measured in kg/litre) or the *bulk density* (measured in kg/m^3) of the energy carriers, their calorific values, can be estimated. Calorific values can change greatly according to their chemical composition and physical state.

Electric energy: the energy of *electricity* is measured generally in *kilowatt hour* (symbol: kWh); it is 3.6 million J (3.6 MJ), since 1 J = 1 Wattsecond which should be multiplied by 3,600 (1 hour = 3,600 s) and 1 k = 1,000. On the other hand, 1 kWh of electric energy is equivalent to 860 kcal of heat; therefore, the conversion factor from the old energy unit calorie to the standard unit of joule can be derived, as follows:

$$1 \text{ kcal} = 3,600/860 = 4.1868 \text{ kJ}$$

The energy equivalent of 1 kWh = 3.6 MJ = 860 kcal relates to the *net (useful) energy* of electricity. But due to the approximately 30% efficiency of generation and transport of electricity energy, some 12,000 kJ equivalent to 2,870 kcal of primary energy sources should be utilized for the generation of 1 kWh of electricity. Hence, when the energy equivalent of electricity is discussed, these two approaches should be clearly distinct.

Attention should be drawn to the fact that the *kilowatthour* is not a unit of power but a unit of energy, and it is equal to the total quantity of energy dissipated when a power of 1 kW operates for a period of 1 hour.

Energy equivalent and conversion factors

In order to compare the energy equivalent of various energy sources and energy carriers having different calorific values and physico-chemical characteristics, various *energy units* and *energy equivalents* are used in the technical-economic calculations.

Although the basic units are *joule* and *kWh*, in practice, however, these are very small, and units and standard prefixes should be used as follows:

1 k	10^3
1 M (mega)	10^6
1 G (giga)	10^9
1 T (tera)	10^{12}
1 P (peta)	10^{15}
1 E (exa)	10^{18}

On the other hand *kWh* relates firstly to electrical energy, and is not a very convenient unit for measuring and expressing the energy content of fossil fuels and solid fuels. Previously, the *kilogram of coal equivalent* (kgCE) was common, but nowadays, the *kilogram of oil equivalent* (kgOE) is more or less the generally accepted unit for the expression of the calorific equivalent of different energy carriers.

Power and efficiency

The *mechanical power* of internal combustion engines was previously measured in horse power (HP), but the new standard unit is *kW* (conversion factor is HP = 0.736 kW).

The *thermal power* of a heat generator was previously expressed in *kcal/h*; the new standard unit is also *kW* (the conversion factor is 1 kW = 860 kcal/h). So, for example: the thermal power of a 100,000 kcal/h boiler is 100,000/860 = 116 kW.

Electric power is still measured in *kW* (being the fraction of a *joule* per second: J/s).

All energy sources and various forms of energy can be transformed to any other forms, but the energy transformation processes are always accompanied by larger or smaller amounts of energy losses, therefore, the *efficiency of energy transformation* largely depends on the type of energy source or carriers, the construction and design of the energy transformation equipment, and the actual operation conditions. Therefore, when the various combustion units and heating systems are compared from energetic and economic points of view, apart from the *specific fuel price* of the various energy carriers expressed in *currency unit per kJ* or *kgOE* (e.g. US$/kJ or US$/kgOE), the *specific net (usable) energy price*, taking into consideration the expected energy efficiency, should be calculated.

Appendix 10.4   Some conversion figures of wood, fuelwood and charcoal

Table 10.1 Conversion figures (air dry, 20% moisture)

1 t wood	1,000 kg
	1.38 m³
	0.343 t oil
	7.33 barrels oil
	90.5 litres kerosene
	3.5 Mkcal
1 t wood (dry basis)	15 GJ
1 t wood (oven dry)	20 GJ
1 m³ wood	0.725 t
	30 lb
1 m³ wood (stacked)	0.276 cord (stacked)
1 cord wood¹	1.25 t (3.62 m³) (general)
1 cord (stacked)	2.12 m³ (solid)
	3.62 m³ 3.62 (stacked)
	128 ft³
1 stere wood[a]	1 m³ (0.725 t)
1 pile wood	0.510 t (510 kg)
1 QUAD	62.5 t wood (od)
	96.2 t wood (green)
1 headload[b]	37 kg
1 t charcoal	6–12 t wood[c]; 30 GJ
1 m³ charcoal	8.28–16.56 m³ wood
	0.250 t
1 kg wet wood	1.2–1.5 m³ producer gas
1 kg dry wood	1.9–2.2 m³ producer gas
1 kg charcoal	4.2–4.7 m³ producer gas
1 t agricultural residues	10–17 GJ[d]

Source: Rosillo-Calle et al. (1996).

Notes
These are approximate figures only and therefore it is important that all conversion factors used are clearly stated.
[a]Cord and stere measures can vary appreciably in actual practice. See Glossary, Appendix 10.1.
[b]The headload varies from place to place. This corresponds to a mature female.
[c]This wide variation is due a number of factors, e.g. species, moisture content, wood density, charcoal piece size, fines, etc.
[d]This large variation is due to moisture content. With a moisture content of 20% the value is 13–15 GJ.

Appendix 10.5   Measuring heating values and moisture contents of biomass

The heating value (HV) of fuels is recorded using two different types of energy content: *gross* and *net*. Although for petroleum the difference between the two is rarely more than about 10%, for biomass fuels with widely varying moisture contents the difference can be very large.

Gross heating value (GHV)

This refers to the total energy that would be released through combustion divided by the weight of the fuel. It is widely used in many countries. It is also called higher heating value (HHV).

Net heating value (NHV)

This refers to the energy that is actually available from combustion after allowing for energy losses from free or combined water evaporation. It is used in all the major international energy statistics. The NHV is always less than GHV, mainly because it does not include two forms of heat energy released during combustion: (1) the energy to vaporize water contained in the fuel and (2) the energy to form water from hydrogen contained in hydrocarbon molecules, and to vaporize it. It is also called lower heating value (LHV). For example:

Combustion process outputs

1   Heat = NHV
2   Hot water vapour formed from hydrogen, including its latent heat of vaporization: fuel + air: combustion = GHV
3   Hot water vapour from contained water, including latent heat
4   Carbon dioxide and monoxide, nitrogen oxides, etc.

Note: 1 = NHV; 1+2+3+4 = GHV.

Furthermore, the difference between NHV and GHV depends largely on the water (and hydrogen) content of the fuel. Petroleum fuels and natural gas contain little water (3–6% or less) but biomass fuels may contain as much as 50–60% water at the point of combustion.

Heating values of biomass fuels are often given as the energy content per unit weight or volume at various stages: *green*, *air-dried* and *oven-dried* material. (See Glossary, Appendix 10.1.)

Most international and national energy statistics are now given in terms of the LHV, in which 1 tonne of oil equivalent is defined as 41.868 GJ. However, some countries and many biomass energy reports and projects, still use HHVs. For fossil fuels such as coal, oil and natural gas, and for most forms of biomass, the LHV is close to 90% of the HHV.

Moisture content

The water contained within biomass material can alter by a factor of 4–5 between initial harvesting (as 'green' crop or wood) and its use as a fuel after some time, during which the material partially dries, or loses water. The water content at any stage in this process, usually termed moisture content and given as a percentage, is measured in two ways, on a 'wet' and a 'dry' basis. Moisture content, dry basis (mcdb) is the weight of water in the biomass divided by the dry weight of biomass. Moisture content, wet basis (mcwb) is the weight of water in the biomass divided by the total weight of the biomass; that is, the weight of water plus the weight of dry biomass.

If the weight of water is W and the weight of dry biomass is D:

$$\mathrm{mcdb} = W/D \quad \mathrm{mcwb} = W/(W + D)$$

To convert between these measures:

$$\mathrm{mcdb} = \mathrm{mcwb}/(1 - \mathrm{mcwb}) \quad \mathrm{mcwb} = \mathrm{mcdb}/(1 + \mathrm{mcdb})$$

The relationship between the two measures is illustrated in the first two columns of Table 10.2 that gives the higher and lower heating values of biomass, in GJ per tonne, according to moisture content and three typical values of the HHV of dry biomass (HHVd); see Kartha et al. (2005), pp. 141–143 for further details.

Table 10.2 Energy content of biomass according to moisture content and use of higher versus lower heating values

Moisture (%)		Higher heating value of dry biomass (HHVd)					
		22 GJ	20 GJ	18 GJ	22 GJ	20 GJ	18 GJ
		HHV and LHV of biomass at given moisture content					
mcwb	*mcdb*	*HHV*	*HHV*	*HHV*	*LHV*	*LHV*	*LHV*
0	0	22.0	20.0	18.0	20.79	18.79	16.79
5	5	20.9	19.0	17.1	19.64	17.74	15.84
10	11	19.8	18.0	16.2	18.49	16.69	14.89
15	18	18.7	17.0	15.3	17.34	15.64	13.94
20	25	17.6	16.0	14.4	16.18	14.58	12.98
25	33	16.5	15.0	13.5	15.03	13.53	12.03
30	43	15.4	14.0	12.6	13.88	12.48	11.08
35	54	14.3	13.0	11.7	12.72	11.42	10.12
40	67	13.2	12.0	10.8	11.57	10.37	9.17
45	82	12.1	11.0	9.90	10.42	9.32	8.22
50	100	11.0	10.0	9.00	9.27	8.27	7.27

Source: See Kartha et al. (2005).

As can be appreciated from the table, moisture has a large effect on the energy content per unit mass of biomass, whereas the use of HHV has a smaller but significant effect. As can be expected, the difference between the lower and higher heating values increases with moisture content (Kartha et al., 2005, pp. 141–143). Table 10.3 illustrates residue product, energy value and moisture content.

Table 10.3 Residue product, energy value and moisture content

Biomass item	Ratio of product: residue	Product energy value (GJt⁻¹)	Product moisture status	Residue energy value (GJt⁻¹)	Residue moisture status
Course grains[a]	1.0:1.3	14.7	20% air dry	13.9	20% air dry
Oats[a]	1.0:1.3	14.7	20% air dry	13.9	20% air dry
Maize[a]	1.0:1.4	14.7	20% air dry	13.0	20% air dry
Sorghum	1.0:1.4	14.7	20% air dry	13.0	20% air dry
Wheat	1.0:1.3	14.7	20% air dry	13.9	20% air dry
Barley	1.0:2.3	14	20% air dry	17.0	Dry weight
Rice	1.0:1.4	14.7	20% air dry	11.7	20% air dry
Sugar cane	1.0:1.6	5.3	48% moisture	7.7	50% moisture
Pulses total[b]	1.0:1.9	14.7	20% air dry	12.8	20% air dry
Dry beans	1.0:1.2	14.7	20% air dry	12.8	20% air dry
Cassava[c]	1.0:0.4	5.6	Harvest	13.1	20% air dry
Potatoes	1.0:0.4	3.6	50% moisture	5.5	60% moisture
Sweet potatoes[c]	1.0:0.4	5.6	Harvest	5.5	Harvest
Fruits	1.0:2.0	3.2	Harvest	13.1	20% air dry
Vegetables	1.0:0.4	3.2	Fresh weight	13.0	20% air dry
Fibre crops[d]	1.0:0.2	18.0	20% air dry	15.9	20% air dry
Seed cotton	1.0:2.1	25.0	Dry weight	25.0	Dry weight
Sunflower[e]	1.0:2.1	25.0	Dry weight	25.0	Dry weight
Soybeans	1.0:2.1	14.7	20% air dry	16.0	20% air dry
Groundnuts[e]	1.0:2.1	25.0	20% air dry	16.0	20% air dry
Tea	1.0:1.2	10.2	20% air dry	13.0	20% air dry
Copra (coconut product)		28.0	5% moisture		
Fibre (coconut residue)	1.0:1.1			16.0	Air dry
Shell (coconut residue)	1.0:0.86			20.0	Air dry

Sources: Hemstock and Hall (1995); Hemstock (2005).

Notes

[a]Values for course grains (including maize, sorghum and oats) are best assumptions based on those values for similar crops given by Ryan and Openshaw, 1991; Senelwa and Hall, 1994; Strehler and Stutzle, 1987; and Woods, 1990.

[b]Values for pulses total are best assumptions based on those values for similar crops given by Ryan and Openshaw, 1991; Senelwa and Hall, 1994; and Strehler and Stutzle, 1987.

[c]Values for cassava and sweet potatoes are best assumptions based on those values for similar crops given by Senelwa and Hall, 1994; and Strehler and Stutzle, 1987.

[d]Values for fibre crops are best assumptions based on those values for similar crops given by Senelwa and Hall, 1994; and Strehler and Stutzle, 1987.

[e]The product: residue ratios for sunflower and groundnuts are best assumptions based on those values for similar crops given by Strehler and Stutzle, 1987.

Appendix 10.6   Measuring sugar and ethanol yields

Sugar yields

On average, for every 100 t of sugar cane ground the following is produced:

- 11.2 t of raw sugar (98.5 Pol)
- 5.0 t of surplus bagasse (49% moisture) = 1,300 kWh surplus electricity (depending on the boiler efficiency)
- 2.7 t of molasses (89 Brix; specific gravity 1.47)
- 3.0 t of filter mud (80% moisture)
- 0.3 t of furnace ash
- 30.0 t of cane top/trash (barbojo)

These quantities are industrial averages and, therefore, vary considerably between countries and between installations within the same country.

Ethanol yields

1 kg invert sugar	0.484 kg ethanol	0.61 litre
1 kg of sucrose	0.510 kg ethanol	0.65 litre
1 kg starch	0.530 kg ethanol	0.68 litre
1 kg ethanol	6,390 kcal	
1 litre ethanol	5,048 kcal	
1 tonne ethanol	7.94 petroleum barrels	
1 tonne ethanol	1,262 litres	
1 tonne ethanol	26.7 GJ (21.1 MJ/litre), LHV; 23.4 GJ/litre, HHV	

Note: maximum theoretical yield of ethanol on a weight basis is 48% of converting glucose.

Sources: Thomas (1985); Biomass Users Network (BUN) figures, http://bioenergy.ornl.gov/papers/misc/energy_conv.html.

Properties of sugar cane

In general, 1 tonne of sugar cane:

- has an energy contents of 1.2 barrels of oil equivalent,
- can produce 118 kg of sugar (+ 10 litres of molasses),
- can produce 85 litres of anhydrous ethanol,
- 89 litres of hydrated ethanol.

Source: UNICA, Brazil (www.portalunica.com.br).

Conversion factors: litre of ethanol to litre of of petrol

The conversion factor is not that simple because it is largely dependent not only on the calorific value of the fuels (i.e. ethanol and petrol) but on the engine efficiency, tuning, blending, etc. For example, while the calorific value of ethanol is much lower, when it burns it does so with much greater efficiency than petrol, thus producing more power, partially offsetting its lower calorific value. In Brazil for example, where ethanol is blended with petrol in a proportion of 24%,[1] the conversion factor varies between 1.2 and 1.3 litres of ethanol/litre of petrol, to take into account the lower heating value and the higher efficiency of ethanol in engines.

This conversion factor is even more complicated in the case of Flex-Fuel vehicles because the blending proportion can vary considerably (i.e. in Brazil from a minimum of 20% to almost 100% blend, although a 50/50% ethanol/petrol blend is currently common). But the complication does not end here because the engine characteristics of Flex-Fuel vehicles are different because the compression ratio of these engines varies; e.g. currently this is between that of a petrol and an ethanol fuel engine (see Rosillo-Calle and Walter, 2006 for further details).

Sugar and ethanol production from sugar cane

The flow diagram in Figure 10.1 illustrates in simplified form the production of sugar and ethanol from sugar cane in Brazil.

Ethanol agro-industrial processes

Table 10.4 details the energy flows involved in ethanol agro-industrial processes in Brazil.

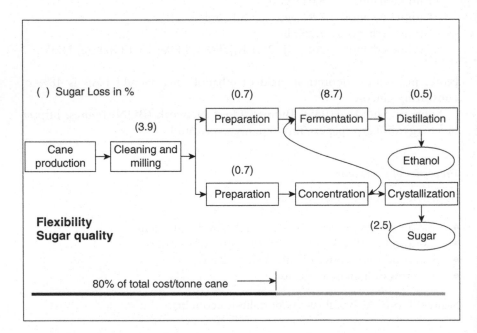

Figure 10.1 Simplified flow diagram for sugar and ethanol production from sugar cane, Brazil. Source: Macedo (2003).

Table 10.4 Energy balance of ethanol agro-industrial process in São Paulo, Brazil

	Energy flows (MJ/tonne of sugar cane)			
	Average (2005/2006)		Projection for 2020	
	Consumption	Production	Consumption	Production
Agricultural activities	210		238	
Industrial process	24		24	
Ethanol production		1,926		2,060
Bagasse surplus		176		0
Electricity surplus[a]		83		972
Total	234	2,185	262	3,032
Energy output/input	9.3		11.6	

Source: Modified from Macedo et al. (2008).

Note

[a]The values for electricity surplus are 9.2 and 135 kWh/tonne of sugar cane for 2005/2006 and 2020, respectively. Considered thermal-electricity equivalences were 9 MJ/kWh (2005) and 7.2 MJ/kWh (2020).

Note

1 This can vary, but in most cases it stays at between 20 and 25% blend.

Appendix 10.7   Carbon content of fossil fuels and bioenergy feedstocks

Carbon content of fossil fuels

Fuel	Carbon content
1 kg of coal	0747 kg
1 tonne of coal	747 kg
1 litre of oil (average)	0.63852 kg
1 tonne of oil	90542 kg (regular)[a]
1 litre of diesel	0.7308 kg
1 tonne of diesel fuel	847 kg[b]
1 m³ of natural gas	0.49 kg

Notes
[a]1418 l/t at 30°C;
[b]1159 l/t at 30°C.

Carbon content of bioenergy feedstocks (approximate)

| Woody crops | 50% |
| Graminaceous (grass and agricultural residues) | 45% |

Sources: http://bioenergy.ornl.gov/papers/misc/energy_conv.html

References and further reading for Appendices 10.1–10.7

BIALY, J. (1979) *Measurement of energy released in the combustion of fuels.* School of Engineering Sciences, Edinburgh University, Edinburgh.

BIALY, J. (1986) *A new approach to domestic fuelwood conservation: guidelines for research.* FAO, Rome.

HEMSTOCK, S.L. (2005) *Biomass energy potential in Tuvalu.* Alofa Tuvalu. Government of Tuvalu Report.

HEMSTOCK, S.L. AND HALL, D.O. (1995). Biomass energy flows in Zimbabwe. *Biomass and Bioenergy* 8, 151–173.

KARTHA, S., LEACH, G. AND RJAN, S.C. (2005) *Advancing bioenergy for sustainable development: guidelines for policymakers and investors.* Energy Sector Management Assistance Programme (ESMAP) Report 300/05. World Bank, Washington DC.

LEACH, G. AND GOWEN, M. (1987) *Household energy handbook: an interim guide and reference manual.* World Bank technical paper no. 67. World Bank, Washington DC.

MACEDO, I.C. (2003) *Technology: key to sustainability and profitability: a Brazilian view.* Cebu, Workshop Technology and Profitability in Sugar Production, International Sugar Council, Philippines, 27 May.

MACEDO, I.C., SEABRA. J.E.A. AND SILVA, J.E.A.R. (2008) Greenhouse gas emissions in the production and use of etanol from sugarcane in Brazil: the 2005/2006 averages and prediction for 2020. *Biomass and Bioenergy* 32, 582–595.

OPENSHAW, K. (1983) Measuring fuelwood and charcoal. In *Wood fuel surveys.* FAO, Rome, pp. 173–178.

ROSILLO-CALLE, F. AND WALTER, A.S. (2006) Global market for bioethanol: historical trends and future prospects. *Energy for Sustainable Development,* Special Issue 10(1), 20–32.

ROSILLO-CALLE, F., FURTADO, P., REZENDE, M.E.A. AND HALL, D.O. (1996) *The charcoal dilemma: finding sustainable solutions for Brazilian industry.* Intermediate Technology Publications, London.

RYAN, P. AND OPENSHAW, K. (1991) *Assessment of bioenergy resources: a discussion of its needs and methodology.* Industry and Energy Development Working Paper, Energy Series Paper no 48. World Bank, Washington DC.

SENELWA, K.A. AND HALL, D.O. (1994) A biomass energy flow chart for Kenya. *Biomass and Bioenergy* 4, 35–48.

STREHLER, A. AND STUTZLE, G. (1987) Biomass residues. In *Biomass: regenerable energy,* D.O. Hall and R.P. Overend (eds). Elsevier Applied Science Publishers, London.

THOMAS, C.Y. (1985) *Sugar: threat or challenge?* IDRC-244e. International Development Research Centre, Ottawa.

WOODS, J. (1990) *Biomass energy flow chart for Zambia – 1988.* King's College London, Division of Biosphere Sciences, London.

The following websites will provide you with a lot of additional information.

http://bioenergy.ornl.gov/papers/misc/energy_conv.html (energy units, bioenergy, units converter)

www.convertit.com (electronic unit converter)

www.eere.energy.gov/biomass/feedstock_databases.html (conversion factors)

www.exe.ac.uk/trol/dictunit/ (a general dictionary of units)

www.fao.org//doccrep/007/ (bioenergy terminology, parameters, units and conversion factors, properties of biofuels: moisture content, energy content, mass, volume and density)

www.portalunica.com.br (information on Brazil sugar cane energy)

Index